"This fertile, innovative text is a new interpretation of the t
that tourism has on place and that place has on tourism. It is a re-imagination
of place through art and science, and therefore, should be essential reading for
anyone interested in seeing how this interdisciplinary concept has been pushed
into new touristic frontiers."

David A. Fennell, *Professor, Dept of Geography & Tourism Studies,*
Brock University, Canada

"*Tourism Interventions: Making or Breaking Places* goes beyond simply exposing the benefits and costs of tourism growth and development by documenting
some of the ways in which interventions by any of tourism's key stakeholders
enhance the positive impacts while alleviating its negative ones on destinations.
Broadly understood as a policy, a strategy, action, event, business idea, collaboration, or partnership, interventions are not always successful, and these rarely
discussed failures are also captured through case studies."

Marion Joppe, *Distinguished Professor Emerita, School of Hospitality,*
Food and Tourism Management, University of Guelph, Canada

Tourism Interventions

This book brings together in one volume, the various types of interventions that can steer tourism towards positive impacts (and/or prevent negative impacts) on the destinations where tourism is taking place.

Interventions in tourism studies have been viewed primarily as 'public interventions' and mainly in the sphere of public policies, planning, and development. This book, however, adopts a larger viewpoint by considering the concept of intervention in areas other than the public sector. The purpose, therefore, is to look into different meanings and uses of the notion of intervention which might involve the initiatives of a variety of actors or agents across locales, borders or scales, as well as how the impacts of tourism on a place have been dealt with. To this end, the book examines tourism interventions and their role in making or breaking places, as initiated and implemented by a variety of stakeholders (public/private sector, NGOs and local communities), by exploring the realities of tourism interventions and how they are utilized to alleviate the negative impacts of tourism; innovative and successful interventions that have contributed to tourism's making of places; and the way in which certain interventions have not been particularly successful or 'failing forward'. This significant volume moves beyond considerations of 'just' policy or 'just' marketing, and brings together different forms of action or inaction in one category, which is a useful response to the variety of actors and initiatives in the tourism space.

This book provides students, researchers, and academics with new insight and understanding of how best to sustainably develop, promote, and manage tourism, and how to help destinations become more resilient in the face of future crises.

Rami K. Isaac is Senior Lecturer in Tourism at the Academy for Tourism, member of the Research Group Tourism Impacts on Society, Breda University of Applied Sciences, The Netherlands and Vice-President of the Research Committee 50 on International Tourism, International Sociologist Association ISA (2014–2025).

Jeroen Nawijn is Senior Lecturer in Tourism at the Academy for Tourism, member of the Research Group Tourism Impacts on Society, Breda University of Applied Sciences, The Netherlands.

Jelena Farkić is Lecturer in Tourism at the Academy for Tourism, member of the Research Group Tourism Impacts on Society Breda University of Applied Sciences, The Netherlands.

Jeroen Klijs is Professor of Tourism at the Academy for Tourism, Research Group Tourism Impacts on Society, Breda University of Applied Sciences, The Netherlands and leads the Research Group Tourism Impacts on Society.

Contemporary Geographies of Leisure, Tourism and Mobility

Series Editor: C. Michael Hall, *Professor at the Department of Management, College of Business and Economics, University of Canterbury, Christchurch, New Zealand*

The aim of this series is to explore and communicate the intersections and relationships between leisure, tourism and human mobility within the social sciences.

It will incorporate both traditional and new perspectives on leisure and tourism from contemporary geography, e.g. notions of identity, representation and culture, while also providing for perspectives from cognate areas such as anthropology, cultural studies, gastronomy and food studies, marketing, policy studies and political economy, regional and urban planning, and sociology, within the development of an integrated field of leisure and tourism studies.

Also, increasingly, tourism and leisure are regarded as steps in a continuum of human mobility. Inclusion of mobility in the series offers the prospect to examine the relationship between tourism and migration, the sojourner, educational travel, and second home and retirement travel phenomena.

The series comprises two strands:

Contemporary Geographies of Leisure, Tourism and Mobility aims to address the needs of students and academics, and the titles will be published simultaneously in hardback and paperback. Titles include:

Literary Fiction Tourism
Understanding the Practice of Fiction-Inspired Travel
Nicola E. MacLeod

Media, Place and Tourism
A World of Imagination
Edited by Stijn Reijnders, Emiel Martens, Deborah Castro, Débora Ribeiro Póvoa, Apoorva Nanjangud, and Rosa Schiavone

Tourism Interventions
Making or Breaking Places
Edited by Rami K. Isaac, Jeroen Nawijn, Jelena Farkić and Jeroen Klijs

For more information about this series, please visit: www.routledge.com/
Contemporary-Geographies-of-Leisure-Tourism-and-Mobility/book-series/SE0522

Tourism Interventions

Making or Breaking Places

Edited by
Rami K. Isaac, Jeroen Nawijn,
Jelena Farkić and Jeroen Klijs

Routledge
Taylor & Francis Group

LONDON AND NEW YORK

First published 2024
by Routledge
4 Park Square, Milton Park, Abingdon, Oxon OX14 4RN

and by Routledge
605 Third Avenue, New York, NY 10158

Routledge is an imprint of the Taylor & Francis Group, an informa business

© 2024 selection and editorial matter, Rami K. Isaac, Jeroen Nawijn, Jelena Farkić and Jeroen Klijs; individual chapters, the contributors

British Library Cataloguing-in-Publication Data
A catalogue record for this book is available from the British Library

ISBN: 978-1-032-58211-5 (hbk)
ISBN: 978-1-032-58212-2 (pbk)
ISBN: 978-1-003-44902-7 (ebk)

DOI: 10.4324/9781003449027

Typeset in Times New Roman
by codeMantra

Contents

Figures

Tables

Contributors

Alberto Amore, University of Oulu, Finland

Viola Ammesdoerfer, University of The Balearic Islands, Spain

Martine Bakker, University of Wageningen, The Netherlands

Bertine Bargeman, Breda University of Applied Sciences, The Netherlands

Marcel Bastiaansen, Breda University of Applied Sciences, The Netherlands

Ella Björn, Faculty of Art and Design, University of Lapland

Wilco Boode, Breda University of Applied Sciences, The Netherlands

Josefien Boor, Breda University of Applied Sciences, The Netherlands

Marc Boumeester, ArtEZ University of the Arts, The Netherlands

Emma Braam, Breda University of Applied Sciences, The Netherlands

Marisa P. de Brito, Breda University of Applied Sciences, The Netherlands

Richard W. Butler, Emeritus Professor, Strathclyde University, Glasgow and visiting professor Breda University of Applied Sciences, The Netherlands

Licia Calvi, Breda University of Applied Sciences, The Netherlands

Deepak Chhabra, Arizona State University, USA

Bartolomé Deyà-Tortella, University of The Balearic Islands, Spain

Lotte van Esch, Breda University of Applied Sciences, The Netherlands

Jelena Farkić, Breda University of Applied Sciences, The Netherlands

Roos Gerritsma, Inholland University of Applied Sciences, The Netherlands

C. Michael Hall, University of Canterbury, New Zealand

Donagh Horgan, Inholland, University of Applied Sciences

Moniek Hover, Breda University of Applied Sciences, The Netherlands

Edward Huijbens, University of Wageningen, The Netherlands

Rami K. Isaac, Breda University of Applied Sciences, The Netherlands

Eunhye Grace Kim, Clemson University, South Carolina, USA

Jeroen Klijs, Breda University of Applied Sciences, The Netherlands

Clio Lambrechts, Visit Flanders, Belgium

Sofía López-Rodríguez, University of The Balearic Islands, Spain

Satu Miettinen, Faculty of Art and Design, University of Lapland

Ondrej Mitas, Breda University of Applied Sciences, The Netherlands

Simone Moretti, Breda University of Applied Sciences, The Netherlands

Jeroen Nawijn, Breda University of Applied Sciences, The Netherlands

Bart Neuts, KU, University of Leuven, Belgium

Vincent Nijs, Visit Flanders, Belgium

Can-Seng Ooi, University of Tasmania, Australia

Karin Peters, University of Wageningen, The Netherlands

Jan van Praet, Visit Flanders, Belgium

Rucitarahma Ristiawan, University of Wageningen, The Netherlands

Tina Šegota, University of Maribor, Slovenia

Alberte Tøttenborg, University of Tasmania, Australia

Steven Valcke, Visit Flanders, Belgium

Juriaan van Waalwijk, Breda University of Applied Sciences, The Netherlands

Kristel Zegers, Breda University of Applied Sciences, The Netherlands

Acknowledgements

The editors have a number of acknowledgements they would like to make to those who have contributed to the development of this volume. We would like first to thank all contributors to this book. We very much appreciate their efforts, time, assistance, understanding, and patience at times with our requests for details and adjustments. Their breadth of viewpoint as well as stimulating and detailed knowledge of their very different subject areas has provided us with unique and wide-ranging topics on the subject of this edited volume. We also would like thank the staff at Routledge for their patience and support throughout the preparation of the book proposal and submission of the manuscript, and in particular Emma Travis and Harriet Cunningham for their continued encouragement and support.

1 Introduction

*Rami K. Isaac, Jeroen Nawijn, Jelena Farkić,
and Jeroen Klijs*

Tourism often takes place in fragile communities, and it has become obvious that in some circumstances, the economic gains might be overshadowed by detrimental implications on the society and the environment that were never anticipated (Dwyer et al., 2004: 307–308; Archer et al., 2005: 79–80). A rising number of scholars and researchers have discovered evidence of both negative and positive impacts of tourism in host regions (Postma & Schmuecker, 2017). Tourism is known and blamed for a variety of socio-economic issues at locations, including crowding, crime, and noise pollution. On the other hand, tourism is touted as a promising means of creating jobs, meaningful encounters, and adding value to inhabitants' overall quality of life. In other words, tourism can *make or break* a place. Notwithstanding the raft of literature exploring both the positive and negative impacts of tourism on destinations globally, few studies have been conducted on the various types of interventions and the ways in which these can steer tourism towards positive impacts (and/or prevent negative impacts) on the destinations where tourism is taking place. These types of interventions continue to be underrepresented in tourism research, and there is currently no book dedicated to them. The focus on interventions is therefore timely, and this book is poised to make a valuable contribution to the existing tourism literature.

Despite the fact that sustainable development has dominated academic and professional debate in tourism for decades, empirical evidence indicates that tourism is actually less sustainable than ever (Hall, 2011, 2019; Rutty et al., 2015; Gössling, et al., 2020; Scott et al., 2016). A number of tourism destinations, such as Barcelona, Venice, and Dubrovnik, experienced a significant increase in tourism pressure in the decades preceding the COVID-19 pandemic. This has consequently led to discussions on the conflict between the positive impacts of having a vibrant tourism sector as a driver of socio-economic development and the negative impacts associated with overtourism, overcrowding, and unbalanced tourism development (Dodds & Butler, 2019a, 2019b). To that end, a growing number of academics, professionals, and policymakers have been discussing the urgency and actions to mitigate the negative socio-environmental impacts of tourism and the consequences of overtourism (Goodwin, 2017; Milano et al., 2019; Peeters et al., 2018; Perkumienė & Pranskūnienė, 2019), and even suggesting an alternative governance paradigm for tourism destinations (Matteucci et al., 2021). Even preceding the pandemic,

DOI: 10.4324/9781003449027-1

which forced the travel industry to a halt, academics were calling for a shift to-wards degrowth strategies, particularly in the case of cities affected by excessive growth and the negative impacts of tourism (Romagosa, 2020).

The main theoretical contributions regarding interventions and tourism have long been concerned with the function of local or national government bodies. Decades ago, Pearce (1998) investigated how public interventions promote the rise of tourism in major cities such as Paris, drawing a link between tourism and urban development. Blake and Sinclair (2007) addressed the economic rationale for the government to get involved in the tourism sector, noting that in case of market failures or severe externalities, the private sector's provision of tourism services would be economically inefficient. Similarly, Smeral (2006) explored the circumstances that support a public-state engagement in tourism promotion, while Bramwell (2010) and Ribarić and Ribarić (2013) interrogated the role of govern-ment intervention in promoting sustainable tourism development. Other authors, such as Abram (2010), analysed the prospect of anthropological interventions in tourism regulations and the notion of intervention is utilized to discuss the role and potential usage of other fields of study within tourism.

As a result, interventions in tourism studies have been viewed primarily as 'pub-lic interventions' and mainly in the sphere of public policies, planning, and devel-opment. Taking a step back is necessary to establish a construct that better fits the purpose of this edited volume, adopt a wider viewpoint, and consider the concept of intervention in areas other than the public sector (Moretti, 2021). The purpose is to look into different meanings and uses of the notion of intervention which might involve the initiatives of a variety of actors or agents across locales, borders, and scales, as well as how the impacts of tourism on a place have been dealt with. In an ideal world, the definition would enable the practical use of the concept of interventions in tourism studies, as well as its contribution to the long-term devel-opment of cities and regions. Following this logic, it appears realistic and rational to investigate generic interpretations of the concept of intervention within social sciences (Moretti, 2021).

Literature offers some useful conceptualizations of interventions. While Midg-ley (2000: 113) defines 'intervention' as a 'purposeful action by an agent to create change', Weiss (2000: 81) states that

> social science research can inform the complex and uncertain work of mov-ing toward effective social change on three levels: by describing and analys-ing the problems that practitioners confront, by identifying better outcomes for individuals and society, and by illuminating strategies of intervention that practitioners can use to move toward better outcomes.

According to Schensul (2009: 241),

> interventions are systematically planned, conducted and evaluated social science-based cultural products intercepting the lives of people and institu-tions in the context of multiple additional events and processes (which also

may be referred to as interventions) that may speed, slow or reduce change towards the desired outcome.

Using and interpreting the contributions above, we adopt Moretti's (2021: 5) definition of a 'tourism intervention' as

> a purposeful action planned and carried out by public institutions, NGOs, private organizations, local community actors, and individuals, or any form of collaboration/partnership among them, that, in the complex framework of tourism management, either proved to contribute or was designed to contribute (or is still designed to contribute, if ongoing) to the socio-cultural landscape.

Following this, an intervention can be a policy, but it can also be a strategy, action, event, business idea, collaboration, or partnership. To this end, the edited volume, across its three main parts, examines tourism interventions and their role in making or breaking places as initiated and implemented by a variety of stakeholders (public/private sector, NGOs, and local communities).

Structure of the Book

Part I includes a number of conceptual and empirical chapters, which play with the concept, practices, and realities of tourism interventions and how they are utilized to optimize the impacts of tourism. These provide an overview of their importance for destination development and preservation of the places in which tourism occurs. Part I also addresses the making of tourism places and the economic and social significance of tourism, but more importantly, the realities of management and marketing, as well as the contributions of various stakeholders in addressing the impacts of tourism at various tourism destinations.

In Chapter 2, Ammesdörfer, Deyá-Tortella, and López-Rodríguez, conducted a systematic literature review of 29 articles to summarize the state of the art of behavioral interventions in the field of tourism. The findings reveal that decision information was the most used behavioral intervention, followed by decision assistance and decision structure. The unique domain resulting was sustainability. The authors identified research gaps, such as the lack of unified categorization and the difficulty of transferring and generalizing the effect of behavioral interventions from one research subject to another. This is followed by Chapter 3 by Boumeester, who theorized and exemplified both imaging and imagination as part of tourism intervention, drawing on several artistic research projects in (shared and touristic use of) the public domain based on 'intensive thinking', which is shorthand for art-based research that centres somaesthetic perception as a gateway to what Gregory Bateson names 'a difference which makes a difference'. The author examined how images of anticipation and expectation contribute to the effects of actualization themselves, set against an unprecedented growth in the production and distribution of images that comes with an indefinite and profound impact on our sense of reality. Aside from the actualized visual output, *imaging* also produces (by proxy) significant expectations of a certain spatial cognizance by the pre-formation of affective angles, frames, and vantage points. The specific intervention discussed by Butler, in Chapter 4, fits several of the definitions cited in this volume, the most appropriate one probably being that of Midgley (2000: 113) who defined an intervention as a 'purposeful action by an agent to create change'. In this case, the intervention was both purposeful and designed to create change, involving the creation of an establishment with a goal of ensuring the stability and viability of an island population, in part through developing tourism. In turn, the increased visitation has encouraged the survival and re-emergence of several traditional is-land crafts, most noticeably Fair Isle knitting, as well as providing support for the island-based ferry boat and subsidized air transport service. The interventions have proved successful in economic, social-cultural, and environmental aspects during their seven decades of existence. Chapter 5, by Ristiawan, Huijbens, and Peters discusses the transformation of places into tourism destinations by analysing the relationship between UNESCO geopark development in Indonesia at the local and regional level and the resulting processes of landscape commodification, highlight-ing the negotiated dimensions of land valuing. The chapter unravels who benefits from the establishment of geoparks, how these interventions rely on creating a rent gap, and how they contribute to developing a tourism destination and the concur-rent transformations of the community. Landscapes are continually transformed and commodified for tourism purposes globally. Indicative of this are transport and service infrastructure investments and real-estate developments with concomitant price increases for land. On the positive side, these investments and developments create opportunities for local and national governments to generate revenues from landscape conservation. At the same time, local communities adapt to the situation by redefining the value of landscapes, which in turn affects their everyday liveli-hood strategies in these landscapes. The Ciletuh Geopark and the Gunung Sewu Geopark are the subjects of this chapter.

Part II contains chapters and case studies on innovative and successful inter-ventions that have contributed to our understanding of the diverse ways in which tourism can make places. These chapters and case studies are based on research programmes and labs that have been sponsored as well as projects. They also un-derline their environmental, economic, and societal values, and analyse the specifi-cities, practicalities, and processes of their implementation, aims, and managerial consequences. These chapters consider what these interventions actually *mean* and how they contribute to the development of a destination, community, society, or humanity.

Chapter 6, by Björn and Miettinen, investigates placemaking through perfor-mance, both in a natural environment and through embodied experience in Finnish Lapland. It discusses place-based arts and service design that have adopted a deep consideration for 'place ethic', which demands consideration and respect for any place that reaches further than merely drawing aesthetic inspiration derived from what is known as the 'tourist gaze'. The chapter asks whether artistic cartogra-phies, such as in modern cartography – that is, the studying of new materialities and ways to present oneself – can consider the cultural value of a place, acknowledging that it is deeply rooted in cultural or natural meanings and traditions that are often rendered invisible or silent. Furthermore, it determines whether arts- and service-design-based placemaking can be utilized to present a critical view on traditional placemaking approaches, which seem to employ locals and address their stories and ways of being, but which still acknowledge colonial conceptualizations of the 'site'. Mitas, Van Waalwijk, Bargeman, Calvi, van Esch, Hover, Boode, and Bastiaansen in Chapter 7, assessed an example of a museum experience aimed at both tourists and residents enhanced by storytelling at the Markiezenhof, a medieval palace in Bergen op Zoom, the Netherlands. Though widely applied by destinations, storytelling methods in experience design are under-researched and not universally validated to provide benefits. In their chapter, the authors describe how a museum exhibit is redesigned according to storytelling principles, then track the emotions of visitors to the exhibit using physiological and self-report methods. Their emotions are compared to emotional models of storytelling to assess the effectiveness of the intervention in creating a quality experience that fosters positive evaluation, intent to recommend, and actual recommendation behavior. They found that visitor emotions were characterized by rising emotional tension, confirming that the experience follows the intended emotional profile of effective storytelling.

Moretti, in Chapter 8, argues that well-designed tourism interventions facilitate the accumulation of different forms of capital within communities that are hosting visitors and those capitals serve as contributors in building up the resilience of tourism destinations and their communities. This strengthens destinations' ability to withstand, cope, or bounce back from social, political, and environmental disruptions. To support this argument, conceptual links between elements of community resilience and Bourdieu's forms of capital have been drawn and further analysed based on the empirical case study of a specific tourism intervention. The discussed approach provides valuable insights for tourism practitioners and governance actors, offering a more comprehensive perspective on the societal value of tourism. It also addresses the increasing number of voices calling for redefining the concept of success in tourism, meaning that tourism impacts should be evaluated beyond the hard numbers of tourist arrivals and the impacts on broader societal values, such as the resilience of tourism destinations and their communities, should be considered.

Chapter 9, by Gerritsma and Horgan, describes a case study in the Netherlands, where the world's first tourism living labs were founded in Amsterdam and Rotterdam. This contribution informs on how the Urban Leisure and Tourism labs have cultivated an environment for open innovation and impactful collaboration, where students are seen as key partners in building an ecosystem from the ground up. It offers high-level descriptions of the route to growing a suitable learning environment with public, private, and community sector partners – and emphasizes the importance of joint problematization and readying a fertile ground for experimentation. The chapter shares opportunities and challenges from the embryonic early years of the labs and showcases some of the most impactful ideas and outputs with transformative potential. It reports on overarching themes for research, including how to embed a regenerative design perspective to bring about social innovation in urban tourism. Sharing critical reflections from the lived experience of leading living lab research, it reports on the complexity of collaborative design and the difficulties of embracing uncertainty and ambiguity in education and practice. Concrete reflections on what *makes or breaks* an intervention are shared at the end of the chapter and touch upon the importance of holding space, continual fresh ideas, long-term presence, and frequent feedback loops. These are in opposition to getting stuck in ideation phases and lack of ownership over outcomes – which can yield resentment in the community.

Chapter 10, by Neuts, Valcke, Lambrechts, Nijs, and Van Praet introduces a pilot study of the destination management organisation (DMO) of the Dutch-speaking Flemish region of Belgium, VISITFLANDERS, and their involvement in a revaluation project of St Godelina's Abbey in Bruges. The case study can be insightful since VISITFLANDERS has followed a strategic and organizational repositioning in line with modern stewardship approaches, reimagining tourism as a means to local well-being and placing local stakeholders central in the tourism development discourse. As part of the strategic operationalization, participatory approaches are followed in larger-scale intervention projects such as the purchase and subsequent reconversion of St Godelina's Abbey. In the summer of 2021, VISITFLANDERS started a participative inquiry in which visitors to the site co-created 4,784 scenarios for future development. These scenarios formed the basis for a content analysis and further exploration by a panel of 40 experts. The intervention is exemplary of a participation process which includes resident and non-expert voices, in the redevelopment of local sites of cultural importance, improving the social sustainability of the conversion project and limiting negative externalities that might arise from top-down processes. While local communities were given a platform for participation, the organization could harvest a variety of heterogeneous and potentially original ideas.

Chapter 11 by de Brito, Calvi, Zegers, Boor, and Braam presented a research study focusing on tourism practices and interventions that cater to the well-being of the locals. This chapter zoomed into the activities happening around mural sites and the importance assigned to them by the local inhabitants of Breda, the Netherlands. Semi-structured interviews, with the use of photo-elicitation, were conducted with the residents. There is evidence that the Blind Walls Gallery as an intervention contributes to feelings of safety and pride among the residents, and triggers dialogue among different users' segments stimulating social interactions among residents and between residents and visitors alike. In addition, murals get intertwined with the sense of place of residents, making murals placemakers, and a source of local identity.

Part III offers a critical analysis of tourism interventions, pointing out which elements have or have not been successful and why this is the case. Failure can result, e.g., from poor enforcement of tourism development policies or the implementation of occasional measures without a clear evidence-based strategy that is informed by a solid knowledge of the underlying system of structural interdependencies (Ramaano, 2021; Butler & Dodds, 2021). The chapters interrogate how diverse actors can work together to create the conditions under which the interventions can help tourism make, rather than break, places. The strength of this part lies in the authors providing clear analyses of interventions and producing knowledge that is an important steppingstone towards future success.

In Chapter 12 Kim and Chhabra present a study on the economic impact of off-highway recreation in the state of Arizona. The purpose was to measure the proximate monetary transactions that influence the incomes of local businesses and employment figures. It aims to understand the overall economic significance of OHV activity in Arizona state. This chapter focuses on specific expenditure allocations and the economic impact of the retained local and out-of-state visitors.

Ooi and Tøttenborg, in Chapter 13, discuss the moral limits of the market. Market logic and mechanisms have come to shape the tourism industry. The first moral limit of the market is when everything, such as tourist attractions, ends up being priced. This would include heritage, nature, and even warm hospitality. If not controlled, this may result in 'repugnant transactions', e.g., allowing unsavory tourist activities in a place of worship. The second moral limit of the market is that the market does not always distribute benefits and costs to the different stakeholders fairly, for instance, hotels may receive tourist revenues, but most locals find tourism to be a nuisance because of overcrowding, inflation, and pollution. Interventions – through guidelines, regulations, taxes, and incentives – in the tourism market are common. Many activities are forbidden, such as hunting in protected parks or entering sacred places in skimpy clothing. Regardless, such interventions have not resolved the many challenges of tourism. Scholars, tourism businesses, and the local community (e.g., NGOs and civil society) have championed ways to mitigate the negative impacts of the tourism market. These interventions include regulating the market through taxes, developing community-based tourism, and encouraging businesses to be better corporate citizens.

Chapter 14, by Bakker, explores the application of the Theory of Change (ToC) in tourism development projects to design more effective and sustainable interventions. ToC helps towards the understanding of the causal relationships between challenges, interventions, and outcomes. Additionally, the chapter delves into various diagnostic tools, including the World Economic Forum's Travel and Tourism Development Index, Destination International's DNEXT Diagnostic Tool, the United Nations Development Programme's Tourism Diagnostics Report, the World Bank's Tourism Diagnostic Toolkit, and Bakker's own Tourism-Driven Inclusive Growth Diagnostic (T-DIGD). Each diagnostic tool has its strengths and weaknesses, making the choice of tool dependent on project size and complexity. By using a diagnostic tool in conjunction with the ToC approach, stakeholders can better understand the challenges and design interventions that lead to an impactful change in projects where the tourism sector is utilized to drive inclusive development.

Chapter 15, by Amore and Hall, provides a critical appraisal on policy issues and policy interventions with direct impacts in tourism as well as policy problems caused by unregulated and market-driven tourism policies. They present a framework illustrating policy failures, and policy-action relationships/policy implementation gaps and non-policies building from evidence collected from different destinations and contexts in developed and developing countries. The chapter then provides an overview and a rethinking of policy action in tourism in opposition to the prevailing mode of governance and public policy making.

Chapter 16, by Šegota, discusses the city of Dubrovnik which has been paraded as 'the Pearl of the Adriatic' since 1979 when endowed with the UNESCO Heritage

Site title. The chapter initially touches upon the importance of culture and heritage as main drivers of Dubrovnik's tourism development. Then, the focus of the chapter shifts towards cruising and film-induced tourism – i.e., the two types of tourism that dominated pre-pandemic decades of the city's tourism development. The focus of the discussion is on (mal)practices in tourism development and promotion on the local and national level that led towards the city becoming a synonym for overtourism. The author presents the Respect the City (RTC) strategic activities implemented pre-pandemic, critically evaluating them, in addition to discussing local and national tourism development (mal)practices post-pandemic. Hence, the critical evaluation of RTC and post-pandemic activities focused on whether the city has been rightfully self-proclaimed as the leader of sustainable and responsible tourism development in the Mediterranean.

Chapter 17, the final chapter, is a closing piece by the editors, who endeavour to create more space for further discussions of interventions in academic and professional contexts following on from this edited volume. It also defines some guidelines for future interventions, discusses the potential of future research, and calls for the revitalization of a critical position in terms of future perspectives on tourism interventions.

References

Abram, S. (2010). Anthropology, tourism and interventions. In J. Scott & T. Selwyn (Eds.), *Thinking Through Tourism* (pp. 231–254). Oxford: Berg.

Archer, B., Cooper, C., & Ruhanen, L. (2005). The positive and negative impacts of tourism. In W. Theobald (Ed.), *Global Tourism* (pp. 87–113). Amsterdam: Elsevier.

Blake, A., & Sinclair, T. (2007). The economic rationale for government intervention in tourism. *Report for the Department for Culture, Media and Sport.*

Bramwell, B. (2010). Participative planning and governance for sustainable tourism. *Tourism Recreation Research,* 35(3), 239–249.

Butler, R., & Dodds, R. (2021). Overcoming overtourism: A review of failure. *Tourism Review,* 77(1), 35–53.

Dodds, R., & Butler, R. (2019a). The phenomena of overtourism: A review. *International Journal of Tourism Cities,* 5(4), 519–528.

Dodds, R., & Butler, R. (2019b). *Overtourism: Issues, realities and solutions.* Berlin, Boston: De Gruyter.

Dwyer, L., Forsyth, P., & Spurr, R. (2004). Evaluating tourism's economic effects: new and old approaches, *Tourism Management,* 25, 307–317.

Goodwin, H. (2017). The challenge of overtourism. In *Responsible Tourism Partnership working paper 4.* Available online at: http://www.millennium-destinations.com/uploads/4/1/9/7/41979675/rtpwp4overtourism012017.pdf (Accessed 22 December 2022).

Gössling, S., Scott, D., & Hall, C. M. (2020). Pandemics, tourism and global change: A rapid assessment of COVID-19. *Journal of Sustainable Tourism,* 29(1), 1–20.

Hall, C. M. (2011). Policy learning and policy failure in sustainable tourism governance: From first- and second-order to third-order change? *Journal of Sustainable Tourism,* 19(4–5), 649–671.

Hall, C. M. (2019). Constructing sustainable tourism development: The 2030 agenda and the managerial ecology of sustainable tourism. *Journal of Sustainable Tourism,* 27(7), 1044–1060

Matteucci, X., Nawijn, J., & Von Zumbusch, J. (2021). A new materialist governance paradigm for tourism destinations. *Journal of Sustainable Tourism*, 30(1), 169–184

Midgley, G. (2000). *Systemic intervention: Philosophy, methodology, and practice*. New York: Kluwer/Plenum.

Milano, C., Novelli, M., & Cheer, J. M. (2019). Overtourism and tourismphobia: A journey through four decades of tourism development, planning and local concerns. *Tourism Planning and Development*, 16(4) 353–357.

Moretti, S. (2021). State of the art of cultural tourism interventions. Deliverable D3.1 of the Horizon 2020 project SmartCulTour (GA number 870708), published on the project website in May 2020: http://www.smartcultour.eu/deliverables (Accessed 21 December 2022).

Pearce, D. G. (1998). Tourism development in Paris: Public intervention. *Annals of Tourism Research*, 25(2), 457–476.

Peeters, P., Gössling, S., Klijs, J., Milano, C., Novelli, M., Dijkmans, C., Eijgelaar, E., Hartman, S., Heslinga, J., Isaac, R., Mitas, O., Moretti, S., Nawijn, J., Papp, B., & Postma, A. (2018). *Research for TRAN Committee – Overtourism: Impact and possible policy responses*.

Perkumienė, D., & Pranskūnienė, R. (2019). Overtourism: Between the right to travel and residents' rights. *Sustainability*, 11(7), 2138.

Postma, A. & Schmuecker, D. (2017). Understanding and overcoming negative impacts of tourism in city destinations: A conceptual model and strategic framework. *Journal of Tourism Futures*, 3(2), 144–156.

Ramaano, A. S. (2021). Tourism policy and environmental impacts in Musina municipality: Lessons from a case study of failure. *Tourism Critique Practice and Theory*, 2(1), 91–114.

Ribarić, I., & Ribarić, H. M. (2013, April). Government intervention in driving the development of sustainable tourism. *2nd International Scientific Conference Tourism in South East Europe 2013*.

Romagosa, F. (2020). The COVID-19 crisis: Opportunities for sustainable and proximity tourism. *Tourism Geographies*, 22(3), 690–694.

Rutty, M., Gössling, S., Scott, D., & Hall, C. M. (2015). The global effects and impacts of tourism: An overview. In C. M. Hall, D. Scott, & S. Gössling (Eds.), *Handbook of tourism and sustainability* (pp. 36–63). N.L.: Routledge.

Schensul, J. J. (2009). Community, culture and sustainability in multilevel dynamic systems intervention science. *American Journal of Community Psychology*, 43(3–4), 241–256.

Scott, D., Hall, C. M., & Gössling, S. (2016). A review of the IPCC Fifth Assessment and implications for tourism sector climate resilience and decarbonization. *Journal of Sustainable Tourism*, 24(1), 8–30.

Smeral, E. (2006). Aspects to justify public tourism promotion: An economic perspective. *Tourism Review*, 61(3), 6–14.

Weiss, J. A. (2000). From research to social improvement: Understanding theories of intervention. *Non-profit and Voluntary Sector Quarterly*, 29(1), 81–110.

Part I

The Concept, Practices, and Realities of Tourism Interventions

2 Promoting Tourists' Responsible Behaviour Through Nudges

A Systematic Literature Review

Viola Ammesdoerfer, Bartolomé Deyà-Tortella, and Sofía López-Rodríguez

Introduction

Tourism grew from 2005 to 2019 with international arrivals up by 81% to 1.47 billion (Statista 2023). Although in 2019 tourism contributed 10.4% to the global GDP (UN Environment Program 2021) it has adverse effects. According to Lenzen et al. (2018) tourism has been responsible for 8% of global greenhouse gas emissions from 2009 to 2013. Forecasts for 2050 project an increase of energy, greenhouse gases, water, and solid waste by 154%, 131%, 152%, and 251%, respectively (UN Environment Program 2021). These significant effects are probably due to the fact that tourists tend to consume more resources on vacation than when they are at home. For example, the study of Gössling and Peeters (2015) estimates that tourists consume more water while they are on vacation compared to when they are at home.

Several instruments have been proposed by academia to mitigate these negative externalities (Pereira-Doel et al. 2019; Deyà-Tortella et al. 2016; Stadelmaier et al. 2022). Although traditionally tourists have been perceived as passive consumers (Dolnicar 2020) within sustainability investigation, there is rapidly growing literature within marketing science on value co-creation behaviors and how customers decide to take action for "mutually valued outcomes through symbiotic relationships" (Jain et al. 2023). Tourists make numerous decisions that impact resource consumption during their stay, and possess the potential to be empowered decision-makers (González Alemán 2018).

In our study, we propose that nudges can effectively impact tourists' decision-making process, and lead to more optimal decisions. The term "nudge" suggests subtly influencing a person's decision-making process to guide them towards a specific choice (Marchiori et al. 2017). Since tourists typically find themselves in pleasurable contexts (Dolnicar et al. 2020), they may resist or reject policies that force certain behaviors, such as banning plastic waste or single-use drinking water bottles (Chatterjee & Barbhuiya 2021). Given this, nudges, being voluntary and requiring minimal cognitive effort (Datta & Mullainathan 2012), could be the ideal approach to influence tourists, who simply intend to enjoy their holidays.

DOI: 10.4324/9781003449027-3

Behavioral Interventions in Tourism

Nudges belong to the broad field of behavioral research and, more precisely, to behavioral economics. As a behavioral change technique, a nudge can be defined as an element of choice architecture. Mele et al. (2021) provide a summary of the earlier concepts by Thaler and Sunstein (2008) as follows: A choice architecture encompasses and impacts external factors that can steer decisions. It organizes the setting, configuration, and presentation of choices and information, thereby influencing human behavior. It involves a choice designer who incorporates subtle prompts (nudges) to influence behavior in a foreseeable manner.

Thaler and Sunstein (2008) define nudges leveraging Kahneman et al.'s (1982) cognitive biases as easy, predictable choice-altering aspects, not mandates, which include small incentives to encourage behavior-related decisions (e.g., retirement savings). Nudges promote better decisions for individuals, the environment, and society (Szaszi et al. 2018). Information alone isn't a nudge or part of choice architecture; it needs design and technique integration.

Previous literature has concluded that nudges effectively impact individual choices in different contexts, like organ donations (Johnson & Goldstein 2003), healthy food selection (Cadario & Chandon 2019), decreasing meat consumption (Hollands et al. 2013), selecting the right insurance (Thaler & Sunstein 2008), increasing vegetable intake at school (Nørnberg et al. 2015), achieving education goals (Damgaard & Nielsen 2018) or combating obesity (Mulderrig 2017).

Prior systematic reviews focused on nudges or behavioral interventions in a generic way, or on their application in specific areas such as energy and water savings, and analysed specific types of nudges (Bergquist et al. 2019; Damgaard & Nielsen 2018; Ropret Homar & Knežević Cvelbar 2021; Szaszi et al. 2018; White et al. 2019). Though, it seems that none of these studies presented an overview of the most effective nudges to be applied in tourism. Nudges in the tourism sector do not have immediate utility for tourists in many cases (White et al. 2019). The studies that focus on nudges in tourism were developed by Souza-Neto et al. (2022) and Nisa et al. (2017) but either did not identify effective nudges or concluded in the specific topic of towel reuse.

This study focuses on tourism-oriented systematic review of behavioral interventions, including nudges. We aimed to provide policymakers, tourism managers, and environmentally oriented organizations with a set of tools for encouraging more environmentally friendly behavior at different points of interaction during a tourist's stay. First, we have provided an overview of nudges and the extant literature. Second, we describe the systematic review process and introduce the intervention categories associated with each category, including subdomains, main effects and any moderators, and synthesize them. Third, after a discussion of the outcome of the systematic literature review, we provide conclusions and suggest recommendations for future research.

Methods

Literature Search

We conducted systematic searches for published articles in two relevant academic databases, Scopus and Web of Science, followed a clear academic research process (Littell et al. 2008), and applied a two-step search strategy similar to that used by Antonova et al. (2021). First, we introduced a key term concerning tourism (touris* OR hotel OR hospitality) in the article title, abstract, and keyword fields. This led to the identification of a total of 408,806 articles. Second, we applied a term concerning behavioral intervention (nudg* OR "choice architecture" OR "behavio* intervention*") to filter the results in the databases (Table 2.1).

Now we identified 609 articles in Scopus and Web of Science after removal of the duplicates. We did not restrict the search by date and filtered in the database by English.

Screening Procedure

After the literature search, we conducted a two-phase screening procedure using pre-defined inclusion and exclusion criteria (Table 2.2). In the first phase, we filtered the articles by title, abstract. and keyword and determined which articles to include and exclude and which should be reviewed again in their entirety later due to doubts about including them in the study. We structured the inclusion and exclusion criteria based on the methodologies of Watson et al. (2017) and Szaszi et al. (2018), which are useful for obtaining unbiased, transparent results and supporting quality assurance (Littell et al. 2008). We included only full text papers and articles from peer-reviewed journals. Furthermore, we only kept studies with a tourism context with empirically investigated experiments on behavioral intervention techniques attributed as a nudge or connected to choice architecture by the articles' authors. Moreover, we only included studies with real behavioral outcomes; therefore, articles that collected data on preferences or attitudes via surveys were excluded. Following this rule, we excluded studies that applied, for instance, Ajzen's (1991) theory of planned behavior (TPB; Coşkun & Yetkin Özbük 2020; Tkaczynski et al. 2020). Although the TPB is a well-established theory, authors like

Table 2.1 Search strings and steps

Step	Theme	Search string	Number of articles (Scopus + Web of Science)	Total
1: search	tourism	(touris* OR hotel OR hospitality)	160,574 + 248,232	408,806
2: search "within"	nudge	(nudg* OR "choice architecture" OR "behavio* intervention*")	576 + 91	609[a]

[a] This number has been cleaned of the 58 duplicate articles.

Table 2.2 Inclusion and exclusion criteria

Inclusion criteria	Exclusion criteria
1. Full-text papers	1. Review articles, conference abstracts, and conference papers
2. Studies that focused on global tourism context and conducted experiments to empirically investigate one or more behavioral intervention techniques attributed as a nudge or connected to choice architecture literature by the original authors	2. Studies that applied interventions restricting the freedom of choice of the target population, including significant economic incentives, or used education, complex decision support systems, or consultation as a choice architecture intervention
3. Studies published in a peer-reviewed journal	
4. Studies written in English (filtered via database search)	
5. Studies that examined behavioral outcome variables (not preferences or attitudes)	

Terry and O'Leary (1995) have criticized the fact that behavioral intentions and actual behaviors differ. According to our exclusion criteria, meta-analyses, literature reviews, and conference papers were not included, as in systematic reviews like those by Goncalves et al. (2020) and Rodrigues et al. (2013).

In the second phase, we read the remaining 46 articles and in some cases, after reviewing the complete paper, we excluded articles that had been included in the first step according to the inclusion and exclusion criteria (Kim et al. 2020; Table 2.2). Thus, 26 articles (24 from Scopus and two from Web of Science) remained in our database. Similar to Watson et al. (2017), we added three articles based on our prior reading, cross-referencing, and snowballing from the database-sourced articles, resulting in 29 articles. No quality ratings were applied because this type of classification might lead to unnecessary bias (Littell et al. 2008; Thompson et al. 2003).

Results

The resulting 29 articles were included in the systematic review (Figure 2.1). These papers were carefully read to identify the main behavioral interventions, domains, effects, moderators, mediators (if present), and outcomes. In five studies, more than one experiment was conducted and analysed. In three studies, no main effect was identified, although behavioral interventions were tested (Babakhani et al. 2020; Eriksson et al. 2019; Starke et al. 2020).

The final articles were quite recent as 27 of the 29 reviewed studies were from the last ten years (2013–2023), and the oldest was from 2008.

Categorization of Behavioral Interventions

Since there is no standardized classification for behavioral interventions (Szaszi et al. 2018) we decided to follow Münscher et al.'s (2016) taxonomy. Thus, we categorized nudges in three groups – decision *information*, *assistance*, and

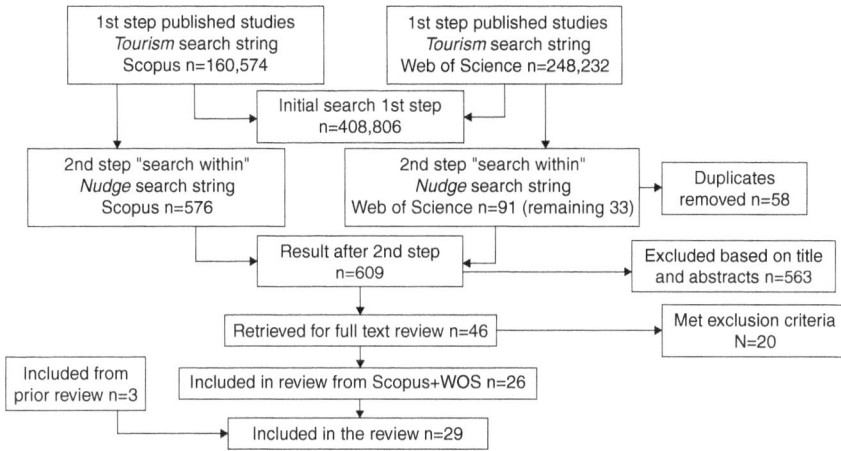

Figure 2.1 Flowchart of the literature search and screening procedure

structure – and nine corresponding techniques. Thaler and Sunstein (2008) initiated the movement with their categorization and the creation of the acronym NUDGE(S) for: offer iNcentives, Understand mapping, Default, Give feedback, Expect errors, and Structure complex choices. Subsequent authors built on this classification (Johnson et al. 2012; Sunstein 2014; Baldwin 2014). Münscher et al. (2016) conducted a broad review of the existing categories, structured them systematically, and put forward a conglomerated taxonomy of nine techniques listed in Table 2.3 (Szaszi et al. 2018).

Table 2.3 shows the number of articles in the present study that used each intervention technique.

We decided to apply Münscher et al.'s (2016) broad taxonomy instead of the one by Schubert (2017) (*caring for the self-image, following the crowd*, and *defaults*) or White et al.'s (2019) SHIFT (Social influence, Habit formation, Individual self, Feelings and cognition, and Tangibility) framework. By adopting Münscher et al.'s (2016) approach, we circumvent narrowing our focus to environmental considerations, thus it was not initially evident that the outcome of our systematic literature review would revolve around sustainability.

Subdomains of Sustainability

All articles in our study represent the domain of sustainability; 28 addressed environmental sustainability, while one addressed social sustainability (Table 2.6). We divided environmental sustainability into subdomains based on environmental issues that were identified by Angelakoglou and Gaidajis (2015). We used their sustainability indicators from different industries, and we found in our database: energy use, human toxicity, ozone depletion, resource consumption, biodiversity, and recycling. According to this categorization, articles about ecological hotel choice would not have a corresponding issue category, so the subdomain *ecological hotel choice* was retained and not categorized further, as suggested by the

Table 2.3 Choice architecture categories and techniques by Münscher et al. (2016) and the number of interventions used in the reviewed articles

Category	Technique	Included in technique (definition)	Interventions in review
A. Decision information	A 1 Translate information	Framing techniques and simplification of available information by translating it into plain language or understandable numerical formats, shifting the focus, reframing, and harnessing loss-aversion.	13
	A 2 Make information visible	Make own/other behavior/information visible via feedback, labels, measurement, conveniently displayed.	6
	A 3 Provide social reference point	Referring to a descriptive norm (what you should do), align one's behavior with others or referring to an opinion leader.	6
B. Decision structure	B 1 Change choice defaults	Opt-in, opt-out techniques, or prompted choice (forcing to actively decide). Set no-action default, use prompted choice.	3
	B 2 Change option-related effort	Increasing/decreasing physical/ financial effort (slightly reducing the barrier to take a decision).	1
	B 3 Change range or composition of options	Changing categories, change grouping of options.	0
	B 4 Change option consequences	Connect decision to benefit/cost, change social consequences of the decision.	0
C. Decision assistance	C 1 Provide reminders	Make information more salient or easier to access.	3
	C 2 Facilitate commitment	Supporting self or public commitment to choice.	3
TOTAL			35

Note: Due to multiple nudging interventions the number of interventions exceeds 29 (the number of articles reviewed).

Organizational Sustainable Performance Index (OSPI) method, where issues can be tailored to meet the specific requirements of the studied industry (Angelakoglou & Gaidajis 2015). We acknowledge some subjectivity in assigning the subdomain to a category, although in cases of doubt, we discussed any concerns and reached an agreement.

Field Experiment Findings on Behavioral Interventions

Table 2.4 provides an overview of the results of the behavioral interventions analysed in this systematic literature review. We sorted the 29 reviewed articles according to the investigated interventions. The first column contains the intervention techniques applied in the studies. Articles that included more than one type of behavioral intervention are included at the end of the table. The third column contains the subdomains of sustainability addressed in each article. In the fourth column, the findings, we provide a short description if necessary for a better comprehension

Table 2.4 Behavioral interventions analysis

Intervention	Author (Year)	Domain	Findings
Information: Translate information	Coghlan (2021)	Wildlife protection (Biodiversity)	Virtual reality game without overt pro-environmental appeal for coral reef donation led to selecting a (turtle) conservation donation as thank you gift (Pro-environmental behavior). Main effect: Information simplification about local challenges via a virtual reality game led to a donation decision.
	Cozzio et al. (2020)	Food choice (Ozone depletion/ Human toxicity)	Informative appeals increased hotel guests' sustainable/ethical food consumption. An appeal of local, organic or fair had more persuasion than a neutral informative appeal. Main effect: Experiential (visual, emotional, participatory) informative appeals led to more ethical food consumption. Moderator: Emotions, participation, visualization.
	Cozzio, Volgger, & Taplin (2021)	Food choice (Ozone depletion/ human toxicity)	Main effect: Factual self-benefit appeals (health), as well as hybrid self/other-benefit appeals (local), led to healthier and more ecological food choice. Moderator: Everyday behaviors (healthy eating, food purchase, environmental behavior), holiday eating attitude.
	Cui et al. (2020)	Ecological hotel choice	Main effect: Physical cleansing increased choice for environmentally friendly hotel. Moderator: N/A; Cultural background and individual or group decision making need to be further investigated. Mediator: Anticipated guilt.
	Nelson et al. (2020)	Wildlife protection (Biodiversity)	Main effect: Valence framing (positive and negative) via plain video and 360° virtual reality increased donation behavior for coral reef protection. 360° virtual reality with negative message was the most effective treatment for one group only (Gili Trawangan). Moderator: Age (all groups), participant's home distance to the sea (only for Gili Trawangan group), income (only for Bogor group).
	Nelson et al. (2021)	Wildlife protection (Biodiversity)	Main effect: 1st experiment both positive and negative framing treatments (sign and/or asking) led to less plastic bag usage in small supermarket. No significant difference between the framed treatments (positive and negative). 2nd experiment exposition to both positive and negatively framed briefings to snorkelers led to mitigate damages to coral reef of snorkeling behavior. Moderators: Previous environmental friendly behavior, amount spent on accomodation (positively correlated with plastic bag use), region of origin.

(Continued)

Table 2.4 Continued

Intervention	Author (Year)	Domain	Findings
	Grazzini et al. (2018)	(Recycling)	Main effect: Loss and gain-framed messages led to increased recycling behavior by hotel guests compared to no messages while loss-framed messages led to higher rates of recycling behavior than gain-framed ones. Moderator: Construal level (concrete/abstract) of the message (the more concrete the message combined with loss framing the more is the effect). Mediator: Perceived self-efficacy.
	Kurz (2018)	Food choice (Ozone depletion)	Main effect: Changing the order of the dishes and putting vegetarian dishes on the front (more salient) led to an increase in vegetarian dishes and reduction of meat consumption in a university canteen, which means less greenhouse gas emission.
	Vinzenz (2019)	Ecological hotel choice	Main effect: The customer online ratings of a hotel generally and its sustainability rating led to a choice of the tourist for a more sustainable hotel. The better a hotel is evaluated, the higher its perceived attractiveness will be. Sustainability rating pictograms create higher trust in sustainability certification of the hotel than sustainability labels.
	Volgger et al. (2021)	Food choice (Ozone depletion/ human toxicity)	Main effect: Health and local (persuasive) messages led to healthier and more ecological food consumption. Moderator: Everyday pro-environmental behaviors (purchasing, eating).
Information: Make information visible	Eriksson et al. (2019)	Food waste (Resource consumption)	Main effect: N/A: Quantification and measurement of food waste in canteens, restaurants, hotels etc. itself are not a guarantee of waste reduction. No behavior intervention led to a significant food waste reduction. Increased quantification time and share of days with a complete recording led to more food waste reduction. Moderator: Automatization of measurement, amount of initial food waste (the higher the initial food waste is, the more is the reduction), duration of measurement period (the longer the measurement period is the more is reduced).
	Babakhani et al. (2020)	Food choice (Ozone depletion)	Labels on carbon footprints should influence the food choice and lead to less environmental-harmful food consumption (chicken burger instead of beef). Local farmer labels do not attract more attention than carbon labels. Main effect: N/A, existence of labels is not significant for food choice. Showing labels with carbon footprint could not be proved to lead to a less harmful food choice.

	[a]Goldstein et al. (2008)	Water consumption (Resource consumption)	Informing hotel guests that other guests generally reused their towels significantly increased towel reuse compared to focusing guests on the importance of environmental protection. Main effect: Provincial norms (comparing with reference persons with similar local settings and circumstances) as a sub-category of descriptive norms (significant vs industry standard) led to more towel reuse by hotel guests. Moderator: Level of perceived similarity among others and a given individual, importance of commonalities for the person.
	Tiefenbeck et al. (2019)	Energy consumption (Energy use)	Main effect: Realtime feedback information of energy consumption led to reduced energy consumption in showers.
Information: Provide social reference point	Gössling et al. (2019)	Water consumption (Resource consumption)	Main effect: A "comprehensive" message containing elements of descriptive social norms based on factual-procedural-effectiveness knowledge, common identities, reciprocity-by-proxy, as well as moral rewards led to a decline in towel and bed linen use for fresh-water-reduction, in comparison to existing in-room messages. Moderator: Nationality, age, length of stay, repeat visits, temperature, hotel standard.
	Morgan & Chompreeda (2015)	Water consumption (Resource consumption)	Main effect: Four types of social norm messages (appeals) led to increased towel reuse of hotel guests, while the injunctive appeal was statistically the most powerful. The other three messages were descriptive plus injunctive, economic (i.e. incentive), and descriptive (their order is reflecting the effectiveness of the message). Moderator: Source markets (domestic or international), family-structure (adults-only vs with children) for the injunctive appeal and the descriptive plus injunctive message.
	Ranson & Guttentag (2019)	[b]Social Sustainability: altruistic guest behavior	Main effect: Human presence of the owner (social presence, human warmth, human sensitivity) led to altruistic cleaning behaviors of the guest in peer-to-peer accomodation.
	Starke et al. (2020)	Energy consumption (Energy use)	Main effect: N/A; Use of descriptive norms (global, similar, experienced norm messages) did not persuade users to choose more energy-saving measures. Mediator: Perceived feasibility.

(Continued)

Table 2.4 Continued

Intervention	Author (Year)	Domain	Findings
Structure: Change choice defaults	Araña et al. (2013)	Greenhouse gas reduction (Ozone depletion)	Main effect: Default messages (opt-in/opt-out) encouraged conference participants to pay for carbon offsetting (local environment fee) for the next conference. Opt-out default frame led to higher probability to pay a carbon offsetting fee than no default. Higher level of WTP in opt-out than in opt-in setting. Moderator: Being a professor/lecturer (increased WTP for carbon offsetting), amount of conference fee (anchoring; increased WTP for carbon offsetting).
	[a] Dolnicar et al. (2019)	Water consumption (Resource consumption/ Ozone depletion)	Main effect: Default with opt-out, called "green" setting (provision of recycled paper serviettes with an option of cotton serviettes), led to the use of recycled paper serviettes instead of thick cotton serviettes for water-saving in a hotel and reduced CO2 emissions.
	Nelson et al. (2019)	Wildlife protection (Biodiversity)	Main effect: Default (vs.open-ended) led to increased donation behavior for coastal conservation. Several suggested donation amounts (anchorpoints), and default opt-in and opt-out at two price levels for coastal conservation (land and sea) were significant and different amounts of donations were achieved for charity organizations. Highest statistical significance was for opt-out. The mean amount donated in opt-out amount at 20 was significantly more than the mean amount donated in anchorpoints. Moderator: Nationality (N-American and Asian spent more than European).
Assistance: Provide reminders	Cozzio, Tokarchuk, & Maurer (2021)	Food waste (Resource consumption)	Persuasive message intervention to reduce plate waste in a hotel buffet setting. Experiential message significantly outperformed a no-message setting. Main effect: Functional, egoistic/nutritional message led to less plate food waste by hotel guests. Moderator: Altruistic message with credible information source like sustainability logo.
	Steckenreuter & Wolf (2013)	Park user fees (Ozone Depletion)	Main effect: Persuasive messages (normative and behavioral) led to increased payment and reduced non-compliance of tourists to pay park user fees. Moderator: Return on investment, benefit thinking, direct attitudes towards Park user fees, local visitor or not (place attachment), ease of processing, motivation to process.

Assistance: Facilitate commitment	Dolnicar et al. (2020)	Food waste (Resource consumption)	Main effect: Information with a flyer or booklet stamp collection game alone or combined with environmental information led to less plate waste on buffet breakfast, so less greenhouse gas emissions. Moderator: Parental principals at home.
Information: Translate information + Assistance: Facilitate commitment + Information: Make information visible	Joo et al. (2018)	Water consumption (Resource consumption)	Main effect: Joint combination of the three and each individual intervention led to increased water-saving by guests in hotel rooms. Interventions were a) social norm message regarding water conservation, b) asking for a non-enforceable voluntary commitment by guests, c) reinforcing the perception of the hotel by showing the hotel's social goal.
Information: Translate information + Structure: Change option-related effort	Kallbekken & Sælen (2013)	Food waste (Resource consumption)	Main effect: Setting (plate size reduction by 3 cm) combined with a message that clients can come back to the buffet as often as they wish led to plate waste reduction at a hotel buffet.
Information: Provide social reference point + Assistance: Provide reminders	[a]Mair & Bergin-Seers (2010)	Water Consumption (Resource Consumption)	Main effect: Information in combination with a request (i.e. reminder) led to highest towel reuse in motels while information only, or combined with a request and a "normative" message, led to increased towel reuse. Moderator: Habits (at home).
Information: Make information visible + Assistance: Facilitate commitment	Baca-Motes et al. (2013)	Water consumption (Resource consumption)	Symbolic and anonymous self-commitment increased towel reuse by hotel guests. Main effect: Specific commitment and self-signalling (via lapel pin) led to reducing water consumption.
Information: Translate information + Provide social reference point	de Visser-Amundson (2020)	Food waste (Resource consumption)	Challenge between restaurants to reduce food waste. Kitchen staff and guests were involved. Priming nudge is the most suitable to combine. Main effect: Combination of appeal by block leader via poster nudge combined with priming intervention via traffic-light smileys and waste weight information led to strongest reduction of food waste followed by commitment nudge or priming nudge in the kitchen in combination with a guest social prompt nudge in the restaurant which led to kitchen food waste reduction and plate waste reduction.

[a] These studies are from a previous manual review.
[b] The only sub-domain of sustainability which is not environmental sustainability.

and we report the main effect of the behavioral intervention and possible moderators and mediators in case there are.

Since five articles applied multiple behavioral interventions, we identified a total of 35 interventions. Three articles reported no *main effect*, so we identified a total of 32 main effects (Table 2.5). The main effect indicated whether a measure (or more measures) was more likely to be chosen in the specific condition than in the baseline (Starke et al. 2020).

As Table 2.5 demonstrates, 22 articles (69% of total) focused on three different decision information techniques: translating information (13 articles), making information visible (4 articles), and providing information as a social reference point (5 articles). Translating involved framing techniques (positive/negative/simplification). Negative framing harnessed loss aversion (Münscher et al. 2016). Negative valence messages increased donations (Nelson et al. 2020). Some studies found no significant difference between positive/negative framing as long as a message was transmitted comprehensively (Coghlan 2021; Nelson et al. 2021).

Volgger et al. (2021) found that personal health and local messages influenced decisions more than neutral information. Tiefenbeck et al. (2019) achieved behavioral change through making information visible (digital meters for energy consumption).

However, Babakhani et al. (2020) and Eriksson et al. (2019) did not realize main effects; neither when a label with carbon footprint information was shown in restaurant menus (Babakhani et al. 2020) nor when the food waste in canteens and restaurants was quantified for a food waste reduction.

Morgan and Chompreeda (2015) influenced behavior using social reference points as decision-inducing information. Ranson and Guttentag (2019) nudged altruistic cleaning behavior. Though, Starke et al. (2020) did not show main effects with descriptive norms.

Six articles (19%) used decision assistance techniques: providing reminders (3 articles) and facilitating commitment (3 articles). Four articles (13%) used decision structure techniques: changing option-related efforts (1 article) and changing choice defaults (3 articles) while default opt-out options were more powerful than default opt-in alternatives (Dolnicar et al. 2019; Nelson et al. 2019; Araña et al. 2013).

Five studies used multiple interventions, like Joo et al. (2018) combining three nudges for towel reuse.

Most interventions targeted tourists or guests, while two food waste studies involved staff (de Visser-Amundson 2020; Eriksson et al. 2019).

Regarding the only identified sustainability domain, it was resource consumption (water and food waste) which was most studied (36%). Ozone depletion, biodiversity, and energy use were also explored. One article studied social sustainability (Ranson and Guttentag 2019).

We identified moderators in 55% of the 29 reviewed studies. Some studies discovered demographic or outside circumstance characteristics like nationality, age, length of stay, temperature, repeat visits, hotel standard, family structure, occupancy, or profession (Nelson et al. 2019; Gössling et al. 2019; Araña et al. 2013).

Table 2.5 Number of main effects for each behavioral intervention

Intervention	Articles With Intervention	Count for frequency	Plus Included in combinations(ᵃ)	Total Interventions	No Main Effect	Total interventions with main effect	%interventions applied	Total	
Information Translate Information	10	10	3	13		13	41%	69%	Information design
Information: Make information visibleᵇ	4	4	2	6	2	4	13%		
Information: Provide social reference pointᵇ	4	4	2	6	1	5	16%		
Assistance: Provide reminders	2	2	1	3		3	9%	19%	Assistance
Assistance: Facilitate commitment	1	1	2	3		3	9%		
Structure: Change choice defaults	3	3	0	3		3	9%	13%	Structure
Structure: Change option-related effort	0	0	1	1		1	3%		
ᵃ*Information: Translate information + Structure: : Change option-related effort*	1	0	0	0		0	0%		Multiple nudging interventions
ᵃ*Information: Translate information + Assistance: Facilitate commitment + Information: Make information visible*	1	0	0	0		0	0%		
ᵃ*Information: Make information visible + Assistance: Facilitate commitment*	1	0	0	0		0	0%		
ᵃ*Information: Translate Information + Provide social reference point*	1	0	0	0		0	0%		
ᵃ*Information: Provide social reference point + Assistance: Provide reminders*	1	0	0	0		0	0%		
*No main effect*ᵇ						3ᵇ			
TOTAL	29	24	11	35	3	32	100%	100%	

Note: There may be discrepancies in the percentages due to rounding.
ᵃCombinations were used to separately count multiple interventions in a single study.
ᵇDenotes interventions with no main effect.

Table 2.6 Most utilized sustainability subdomains in the reviewed articles

Subdomains	Articles of subdomain	Counter of frequency	Plus included in combinations	Total subdomains	% subdomains	TOTAL
Water consumption (Resource consumption)	6	6	1	7	21	Resource Consumption 36
Food waste (Resource consumption)z	5	5	0	5	15	
Food choice (Ozone depletion)	2	2	4	6	18	Ozone depletion 24
Park user fees (Ozone depletion)	1	1	0	1	3	
Greenhouse gas reduction (Ozone depletion)	1	1	0	1	3	
Wildlife protection (Biodiversity)	4	4	0	4	12	
(Human toxicity)	0	0	3	3	9	
Energy consumption (Energy use)	2	2	0	2	6	
Ecological hotel choice	2	2	0	2	6	
(Recycling)	1	1	0	1	3	
Social Sustainability: altruistic guest behavior[a]	1	1	0	1	3	
Food choice (Ozone depletion+Human toxicity)[b]	3	0	0	0	–	
Water consumption (Resource consumption+Ozone depletion)[b]	1	0	0	0	–	
Total general	29	25	8	33	100	

Note: The categories presented by Angelakoglou and Gaidajis (2015) are in parentheses.

[a] Social sustainability was not considered an environmental sustainability category by Angelakoglou and Gaidajis (2015).
[b] If two categories were assigned by the authors, they count in each and are in italics.

Further, pro-environmental everyday behavior reinforced the main effect (Volgger et al. 2021). Endorsement by hotel managers and reinforcement by parental principals were other identified moderators (Cozzio, Volgger, & Taplin 2021; Dolnicar et al. 2020). More diverse moderators included message credibility, emotions, distance to the sea combined with income, and level of automatization (Cozzio, Tokarchuk, & Maurer 2021; Cozzio et al. 2020; Nelson et al. 2020; Eriksson et al. 2019). Message construal level (how difficult is it to understand the message), concrete messages, and loss framing also influenced the impact (Grazzini et al. 2018). Perceived return on investment and similarities with others acted as moderators (Steckenreuter & Wolf 2013; Goldstein et al. 2008).

In three studies we identified mediators such as anticipated guilt, perceived self-efficacy, and perceived feasibility (Cui et al. 2020; Grazzini et al. 2018; Starke et al. 2020).

Hedonism in tourism was not a main effect, moderator, or mediator, but its importance for tourists was noted as a result from a coding exercise supported by MAXQDA software and we recommend it is considered in future studies.

Discussion and Conclusion

The main objective of this systematic literature review is to explore the application of behavioral interventions in the field of tourism. We found that most relevant academic research has been conducted within the last decade. We followed a systematic process which resulted in 29 articles. It was challenging to find field experiments with behavioral interventions leading to actual or intended changes in tourist behavior. The identified articles from the review mainly focused on sustainability. Information design interventions, such as translating information, making it visible, or providing it via a social reference point, were prevalent with positive effects on environmentally friendly behaviors. Three cases did not show significant main effects. The majority of the 32 interventions with main effects employed information design.

Several moderators were identified, including demographic characteristics, roles (e.g., conference attendees), everyday behavior, and endorsement. Additionally, the credibility of an enterprise, such as a hotel, based on labelling, statements, or actions, reinforced the effectiveness of interventions aimed at changing tourist or hotel guest behavior (Cozzio, Tokarchuk, & Maurer 2021).

A critical question that remains unanswered is whether nudges, which belong to the concept of *libertarian paternalism*, are the most effective approach, or if other techniques, such as *boosts* (Congiu & Moscati 2022), should be explored. The potential manipulative nature of nudges has raised concerns, and alternative approaches such as active choice, which empowers decision-makers without boosting specific consequences, should be considered.

The study's limitations included a scarcity of tourism-related field experiments, due to financial and organizational constraints faced by hotel management and staff. The research focused on nudges in tourism, with limited data on non-accommodation providers. Despite these limitations, the study delivers valuable implications, highlighting the potential for environmental sustainability through behavioral interventions in various aspects of hotel management, including food waste reduction, energy conservation, and plastic avoidance. When implementing nudges as well as establishing policy guidelines, researchers, policymakers, and tourism management should consider the hedonic and pleasure-seeking needs of tourists when designing choice architecture (Rodriguez-Sanchez et al. 2020). The study also suggests the possibility of influencing decision-making not only among tourists but also employees, tourism company management, and even policymakers.

Future research is needed to expand the number of real-life tourism-related field experiments to determine the effectiveness of behavioral interventions. We recommend that a complementary experiment will be conducted by applying the insights of the present study to other research fields.

The first contribution of this systematic literature review is that it summarizes the current state of the art of nudge interventions in tourism. Second, it shows that nudges applied in tourism studies have mainly focused on the preservation of the environment instead of social or health-related well-being, community economic development, or other topics. Third, it demonstrates that there is no ideal nudge. Fourth, although there is no clear, simple outcome for an ideal nudge, the most efficient nudges are related to information design. Fifth, it did not identify an empirical experimental study that showed that the exposure to labels led to a main effect on decision-making. Sixth, it demonstrates that research containing tourism-related field experiments about nudges is scarce. Finally, it discusses nudge variants and possible future developments for the application in a tourism context.

References

Ajzen, I. (1991). The theory of planned behavior. *Organizational Behavior and Human Decision Processes, 50*(2), 179–211. https://doi.org/10.1016/0749-5978(91)90020-T.

Angelakoglou, K., & Gaidajis, G. (2015). A review of methods contributing to the assessment of the environmental sustainability of industrial systems. *Journal of Cleaner Production, 108,* 725–747. https://doi.org/10.1016/j.jclepro.2015.06.094.

Antonova, N., Ruiz-Rosa, I., & Mendoza-Jiménez, J. (2021). Water resources in the hotel industry: A systematic literature review. *International Journal of Contemporary Hospitality Management, 33*(2), 628–649. https://doi.org/10.1108/IJCHM-07-2020-0711.

Araña, J. E., & León, C. J. (2016). Are tourists animal spirits? Evidence from a field experiment exploring the use of non-market based interventions advocating sustainable tourism. *Journal of Sustainable Tourism, 24*(3), 430–445. https://doi.org/10.1080/09669582.2015.1101128.

Araña, J. E., León, C. J., Moreno-Gil, S., & Zubiaurre, A. R. (2013). A comparison of tourists' valuation of climate change policy using different pricing frames. *Journal of Travel Research, 52*(1), 82–92. https://doi.org/10.1177/0047287512457260.

Babakhani, N., Lee, A., & Dolnicar, S. (2020). Carbon labels on restaurant menus: do people pay attention to them? *Journal of Sustainable Tourism, 28*(1), 51–68. https://doi.org/10.1080/09669582.2019.1670187.

Baca-Motes, K., Brown, A., Gneezy, A., Keenan, E. A., & Nelson, L. D. (2013). Commitment and behavior change: Evidence from the field. *Journal of Consumer Research*, *39*(5), 1070–1084. https://doi.org/10.1086/667226.

Baldwin, R. (2014). From regulation to behaviour change: Giving nudge the third degree. Retrieved 14 May 2023 from: http://blogs.cabinetoffice.gov.uk/behavioural-insights-team/author/behavioural-insights-team/.

Bergquist, M., Nilsson, A., & Schultz, W. P. (2019). A meta-analysis of field-experiments using social norms to promote pro-environmental behaviors. *Global Environmental Change*, *59*. https://doi.org/10.1016/j.gloenvcha.2019.101941.

Cadario, R., & Chandon, P. (2019). Viewpoint: Effectiveness or consumer acceptance? Tradeoffs in selecting healthy eating nudges. *Food Policy*, *85*(April), 1–6. https://doi.org/10.1016/j.foodpol.2019.04.002.

Chatterjee, D., & Barbhuiya, M. R. (2021). Bottled water usage and willingness to pay among Indian tourists: Visual nudges and the theory of planned behaviour. *Scandinavian Journal of Hospitality and Tourism*, *21*(5), 531–549. https://doi.org/10.1080/15022250.2021.1974544.

Coghlan, A. (2021). Can ecotourism interpretation influence reef protective behaviours? Findings from a quasi-experimental field study involving a virtual reality game. *Journal of Ecotourism*, *21*(2). https://doi.org/10.1080/14724049.2021.1971240.

Congiu, L., & Moscati, I. (2022). A review of nudges: Definitions, justifications, effectiveness. *Journal of Economic Surveys*, 188–213. https://doi.org/10.1111/joes.12453.

Coşkun, A., & Yetkin Özbük, R. M. (2020). What influences consumer food waste behavior in restaurants? An application of the extended theory of planned behavior. *Waste Management*, *117*, 170–178. https://doi.org/10.1016/j.wasman.2020.08.011.

Cozzio, C., Volgger, M., Taplin, R., & Woodside, A. G. (2020). Nurturing tourists' ethical food consumption: Testing the persuasive strengths of alternative messages in a natural hotel setting. *Journal of Business Research*, *117*(August), 268–279. https://doi.org/10.1016/j.jbusres.2020.05.050.

Cozzio, C., Tokarchuk, O., & Maurer, O. (2021). Minimising plate waste at hotel breakfast buffets: An experimental approach through persuasive messages. *British Food Journal*, *123*(9). https://doi.org/10.1108/BFJ-02-2021-0114.

Cozzio, C., Volgger, M., & Taplin, R. (2021). Point-of-consumption interventions to promote virtuous food choices of tourists with self-benefit or other-benefit appeals: A randomised field experiment. *Journal of Sustainable Tourism*, *30*(6), 1301–1319. https://doi.org/10.1080/09669582.2021.1932936.

Cui, Y., Errmann, A., Kim, J., Seo, Y., Xu, Y., & Zhao, F. (2020). Moral effects of physical cleansing and pro-environmental hotel choices. *Journal of Travel Research*, *59*(6), 1105–1118. https://doi.org/10.1177/0047287519872821.

Damgaard, M. T., & Nielsen, H. S. (2018). Nudging in education. *Economics of Education Review*, *64*(March), 313–342. https://doi.org/10.1016/j.econedurev.2018.03.008.

Datta, S., & Mullainathan, S. (2012). Behavioral design: A new approach to development policy. In *CGD Policy Paper* (Issue 16). www.cgdev.org.

de Visser-Amundson, A. (2020). A multi-stakeholder partnership to fight food waste in the hospitality industry: A contribution to the United Nations Sustainable Development Goals 12 and 17. *Journal of Sustainable Tourism*, *2*, 1–28. https://doi.org/10.1080/09669582.2020.1849232.

Deyà-Tortella, B., Garcia, C., Nilsson, W., & Tirado, D. (2016). The effect of the water tariff structures on the water consumption in Mallorcan hotels. *Water Resources Research*, *52*(8), 6386–6403. https://doi.org/10.1002/2016WR018621.

Dolnicar, S. (2020). Designing for more environmentally friendly tourism. *Annals of Tourism Research*, *84*(June), 102933. https://doi.org/10.1016/j.annals.2020.102933.

Dolnicar, S., Juvan, E., & Grün, B. (2020). Reducing the plate waste of families at hotel buffets – A quasi-experimental field study. *Tourism Management*, *80*(March), 104103. https://doi.org/10.1016/j.tourman.2020.104103.

Dolnicar, S., Knežević Cvelbar, L., & Grün, B. (2019). Changing service settings for the environment: How to reduce negative environmental impacts without sacrificing tourist satisfaction. *Annals of Tourism Research*, *76*, 301–304. https://doi.org/10.1016/j.annals.2018.08.003.

Eriksson, M., Malefors, C., Callewaert, P., Hartikainen, H., Pietiläinen, O., & Strid, I. (2019). What gets measured gets managed – or does it? Connection between food waste quantification and food waste reduction in the hospitality sector. *Resources, Conservation and Recycling: X*, *4*(September), 100021. https://doi.org/10.1016/j.rcrx.2019.100021.

Foxall, G. R., Oliveira-Castro, J. M., James, V. K., Yani-de-Soriano, M. M., & Sigurdsson, V. (2006). Consumer behavior analysis and social marketing: The Case of environmental conservation. *Behavior and Social Issues*, *15*(1), 101–125. https://doi.org/10.5210/bsi.v15i1.338.

Goldstein, N. J., Cialdini, R. B., & Griskevicius, V. (2008). A room with a viewpoint: Using social norms to motivate environmental conservation in hotels. *Journal of Consumer Research*, *35*(3), 472–482. https://doi.org/10.1086/586910.

Goncalves, J., Mateus, R., Silvestre, J. D., & Roders, A. P. (2020). Going beyond Good intentions for the sustainable conservation of built heritage: A Systematic literature review. *Sustainability*, *12*(22), 9649. https://doi.org/10.3390/su12229649.

González Alemán, H. (2018). Reflexiones en torno al poder del consumidor alimentario. *Revista de Bioética y Derecho*, *42*, 23–32. https://revistes.ub.edu/index.php/RBD/article/view/19876/23318.

Gössling, S. (2014). New performance indicators for water management in tourism. *Tourism Management*, 46, 233–244. https://doi.org/10.1016/j.tourman.2014.06.018.

Gössling, S., Araña, J. E., & Aguiar-Quintana, J. T. (2019). Towel reuse in hotels: Importance of normative appeal designs. *Tourism Management*, *70*, 273–283. https://doi.org/10.1016/J.TOURMAN.2018.08.027.

Gössling, S., & Peeters, P. (2015). Assessing tourism's global environmental impact 1900–2050. *Journal of Sustainable Tourism*, *23*(5), 639–659. https://doi.org/10.1080/09669582.2015.1008500.

Grazzini, L., Rodrigo, P., Aiello, G., & Viglia, G. (2018). Loss or gain? The role of message framing in hotel guests' recycling behaviour. *Journal of Sustainable Tourism*, *26*(11), 1944–1966. https://doi.org/10.1080/09669582.2018.1526294.

Hollands, G. J., Shemilt, I., Marteau, T. M., Jebb, S. A., Kelly, M. P., Nakamura, R., Suhrcke, M., & Ogilvie, D. (2013). Altering choice architecture to change population health behavior: A large-scale conceptual and empirical scoping review of interventions within micro-environments. *BMC Public Health*, *13*(December), 1–203.

Jain, S., Sharma, K., & Devi, S. (2023). The dynamics of value co-creation behavior: A systematic review and future research agenda. *International Journal of Consumer Studies*, September, 1–23. https://doi.org/10.1111/ijcs.12993.

John, P., Martin, A., & Mikołajczak, G. (2023). Support for behavioral nudges versus alternative policy instruments and their perceived fairness and efficacy. *Regulation & Governance*, *17*(2), 363–371. https://doi.org/10.1111/rego.12460.

Johnson, E. J., & Goldstein, D. (2003). Do defaults save lives? *Science*, *302*(5649), 1338–1339. https://doi.org/10.1126/science.1091721.

Johnson, E. J., Shu, S. B., Dellaert, B. G. C., Fox, C., Goldstein, D. G., Häubl, G., Larrick, R. P., Payne, J. W., Peters, E., Schkade, D., Wansink, B., & Weber, E. U. (2012). Beyond nudges: Tools of a choice architecture. *Marketing Letters*. https://doi.org/10.1007/s11002-012-9186-1.

Joo, H. H., Lee, J., & Park, S. (2018). Every drop counts: A water conservation experiment with hotel guests. *Economic Inquiry*, *56*(3), 1788–1808. https://doi.org/10.1111/ecin.12563.

Kahneman, D., Slovic, P., & Tversky, A. (1982). *Judgment under uncertainty: Heuristics and biases*. Cambridge University Press.

Kahneman, D. (2011). *Thinking, fast and slow*. Farrar, Straus and Giroux.

Kallbekken, S., & Sælen, H. (2013). "Nudging" hotel guests to reduce food waste as a win-win environmental measure. *Economics Letters*, *119*(3), 325–327. https://doi.org/10.1016/j.econlet.2013.03.019.

Kim, J., Kim, S., Lee, J. S., Kim, P. B., & Cui, Y. (2020). Influence of choice architecture on the preference for a pro-environmental hotel. *Journal of Travel Research*, *59*(3), 512–527. https://doi.org/10.1177/0047287519841718.

Kurz, V. (2018). Nudging to reduce meat consumption: Immediate and persistent effects of an intervention at a university restaurant. *Journal of Environmental Economics and Management*, *90*, 317–341. https://doi.org/10.1016/j.jeem.2018.06.005.

Lenzen, M., Sun, Y. Y., Faturay, F., Ting, Y. P., Geschke, A., & Malik, A. (2018). The carbon footprint of global tourism. *Nature Climate Change 2018*, *8*(6), 522–528. https://doi.org/10.1038/s41558-018-0141-x.

Littell, J. H., Corcoran, J., & Pillai, V. (2008). S*ystematic reviews and meta-analysis*. Oxford University Press.

Mair, J., & Bergin-Seers, S. (2010). The effect of interventions on the environmental behaviour of Australian motel guests. *Tourism and Hospitality Research*, *10*(4), 255–268. https://doi.org/10.1057/thr.2010.9.

Marchiori, D. R., Adriaanse, M. A., & De Ridder, D. T. D. (2017). Unresolved questions in nudging research: Putting the psychology back in nudging. *Social and Personality Psychology Compass*, *11*(1), e12297. https://doi.org/10.1111/spc3.12297.

Mele, C., Russo Spena, T., Kaartemo, V., & Marzullo, M. L. (2021). Smart nudging: How cognitive technologies enable choice architectures for value co-creation. *Journal of Business Research*, *129*(August), 949–960. https://doi.org/10.1016/j.jbusres.2020.09.004.

Michie, S., Richardson, M., Johnston, M., Abraham, C., Francis, J., Hardeman, W., Eccles, M. P., Cane, J., & Wood, C. E. (2013). The behavior change technique taxonomy (v1) of 93 hierarchically clustered techniques: Building an international consensus for the reporting of behavior change interventions. *Annals of Behavioral Medicine*, *46*(1), 81–95. https://doi.org/10.1007/s12160-013-9486-6.

Morgan, M., & Chompreeda, K. (2015). The relative effect of message-based appeals to promote water conservation at a tourist resort in the gulf of Thailand. *Environmental Communication*, *9*(1), 20–36. https://doi.org/10.1080/17524032.2014.917689.

Mulderrig, J. (2017). Reframing obesity: A critical discourse analysis of the UK's first social marketing campaign. *Critical Policy Studies*, *11*(4), 455–476. https://doi.org/10.1080/19460171.2016.1191364.

Münscher, R., Vetter, M., & Scheuerle, T. (2016). A review and taxonomy of choice architecture techniques. *Journal of Behavioral Decision Making*, *29*(5), 511–524. https://doi.org/10.1002/bdm.1897.

Nelson, K. M., Anggraini, E., & Schlüter, A. (2020). Virtual reality as a tool for environmental conservation and fundraising. *PLoS ONE*, *15*(4), 1–22. https://doi.org/10.1371/journal.pone.0223631.

Nelson, K. M., Bauer, M. K., & Partelow, S. (2021). Informational nudges to encourage pro-environmental behavior: Examining differences in message framing and human interaction. *Frontiers in Communication, 5*(February), 1–15. https://doi.org/10.3389/fcomm.2020.610186.

Nelson, K. M., Partelow, S., & Schlüter, A. (2019). Nudging tourists to donate for conservation: Experimental evidence on soliciting voluntary contributions for coastal management. *Journal of Environmental Management, 237*(January), 30–43. https://doi.org/10.1016/j.jenvman.2019.02.003.

Nisa, C., Varum, C., & Botelho, A. (2017). Promoting sustainable hotel guest behavior: A Systematic review and meta-analysis. *Cornell Hospitality Quarterly, 58*(4), 354–363. https://doi.org/10.1177/1938965517704371.

Nørnberg, T. R., Houlby, L., Skov, L. R., & Peréz-Cueto, F. J. A. (2015). Choice architecture interventions for increased vegetable intake and behaviour change in a school setting: A systematic review. *Perspectives in Public Health, 136*(3), 132–142. https://doi.org/10.1177/1757913915596017.

Pereira-Doel, P., Font, X., Wyles, K., Pereira-Moliner, J., & Researcher, P. (2019). Showering smartly: A field experiment using water-saving technology to foster pro-environmental behaviour among hotel guests. *E-Review of Tourism Research (ERTR), 17*(3). http://ertr.tamu.eduhttp//ertr.tamu.eduhttp://ertr.tamu.edu.

Pleshcheva, V., & Schmidt, K. M. (2019). Metric and scale effects in consumer preferences for environmental benefits (No. 147; Collaborative Research Center Transregio 190). http://hdl.handle.net/10419/208047.

Ranson, P., & Guttentag, D. (2019). "Please tidy up before leaving": Nudging Airbnb guests toward altruistic behavior. *International Journal of Culture, Tourism, and Hospitality Research, 13*(4), 524–530. https://doi.org/10.1108/IJCTHR-06-2019-0101.

Rodrigues, A., Sniehotta, F. F., & Araujo-Soares, V. (2013). Are interventions to promote sun-protective behaviors in recreational and tourist settings effective? A systematic review with meta-analysis and moderator analysis. *Annals of Behavioral Medicine, 45*(2), 224–238. https://doi.org/10.1007/s12160-012-9444-8.

Rodriguez-Sanchez, C., Sancho-Esper, F., Casado-Díaz, A. B., & Sellers-Rubio, R. (2020). Understanding in-room water conservation behavior: The role of personal normative motives and hedonic motives in a mass tourism destination. *Journal of Destination Marketing and Management, 18*(September), 100496. https://doi.org/10.1016/j.jdmm.2020.100496.

Ropret Homar, A., & Knežević Cvelbar, L. (2021). The effects of framing on environmental decisions: A systematic literature review. *Ecological Economics, 183*(July). https://doi.org/10.1016/j.ecolecon.2021.106950.

Schubert, C. (2017). Green nudges: Do they work? Are they ethical? *Ecological Economics, 132*, 329–342. https://doi.org/10.1016/j.ecolecon.2016.11.009.

Souza-Neto, V., Marques, O., Mayer, V. F., & Lohmann, G. (2022). Lowering the harm of tourist activities : A systematic literature review on nudges. *Journal of Sustainable Tourism, 31*(9), 2173–2194. https://doi.org/10.1080/09669582.2022.2036170.

Stadelmaier, J., Rehfuess, E. A., Forberger, S., Eisele-Metzger, A., Nagavci, B., Schünemann, H. J., Meerpohl, J. J., & Schwingshackl, L. (2022). Using GRADE evidence to decision frameworks to support the process of health policy-making: An example application regarding taxation of sugar-sweetened beverages. *European Journal of Public Health, 32*(4), iv92–iv100. https://doi.org/10.1093/eurpub/ckac077.

Starke, A., Willemsen, M., & Snijders, C. (2020). With a little help from my peers: Depicting social norms in a recommender interface to promote energy conservation. *IUI*

'20: 25th International Conference on Intelligent User Interfaces, 568–578. https://doi.org/10.31219/osf.io/g6q5v.

Statista (2023). Number of international tourist arrivals worldwide 2005–2022 by region. Retrieved 20 July 2023 from: https://www.statista.com/statistics/186743/international-tourist-arrivals-worldwide-by-region-since-2010/.

Steckenreuter, A., & Wolf, I. D. (2013). How to use persuasive communication to encourage visitors to pay park user fees. *Tourism Management, 37,* 58–70. https://doi.org/10.1016/j.tourman.2013.01.010.

Sunstein, C. R. (2014). Why nudge: The politics of libertarian paternalism. In *Why nudge: The politics of libertarian paternalism*. Yale University Press.

Szaszi, B., Palinkas, A., Palfi, B., Szollosi, A., & Aczel, B. (2018). A Systematic scoping review of the choice architecture movement: Toward understanding when and why nudges work. *Journal of Behavioral Decision Making, 31*(3), 355–366. https://doi.org/10.1002/bdm.2035.

Terry, D. J., & O'Leary, J. E. (1995). The theory of planned behaviour: The effects of perceived behavioural control and self-efficacy. *British Journal of Social Psychology, 34,* 199–220. https://doi.org/10.1111/j.2044-8309.1995.tb01058.x.

Thaler, R. H., & Sunstein, C. R. (2008). Nudge: Improving decisions about health, wealth, and happiness. In Leonard, T. C. (Ed.), *Constitutional political economy, 19*(4), 356–360. Yale University Press.

Thompson, J., Deeks, S. G., & Altmann, J. J. (2003). Measuring inconsistency in meta-analyses. *British Medical Journal, 327,* 557–560.

Tiefenbeck, V., Wörner, A., Schöb, S., Fleisch, E., & Staake, T. (2019). Real-time feedback reduces energy consumption among the broader public without financial incentives. *Nature Energy, 4*(October), 831–832. https://doi.org/10.1038/s41560-019-0480-5.

Tkaczynski, A., Rundle-Thiele, S., & Truong, V. D. (2020). Influencing tourists' pro-environmental behaviours: A social marketing application. *Tourism Management Perspectives, 36* (August), 100740. https://doi.org/10.1016/j.tmp.2020.100740.

UN Environment Program (2021). Travel and tourism industry chart new, greener course at COP 26. Retrieved 8 May 2022 from: https://www.unep.org/news-and-stories/story/travel-and-tourism-industry-chart-new-greener-course-cop-26.

Vecchio, R., & Cavallo, C. (2019). Increasing healthy food choices through nudges: A systematic review. *Food Quality and Preference, 78*(May), 103714. https://doi.org/10.1016/j.foodqual.2019.05.014.

Vinzenz, F. (2019). The added value of rating pictograms for sustainable hotels in classified ads. *Tourism Management Perspectives, 29*(October), 56–65. https://doi.org/10.1016/j.tmp.2018.10.006.

Volgger, M., Cozzio, C., & Taplin, R. (2021). What drives persuasion to choose healthy and ecological food at hotel buffets: Message, receiver or sender? *Asia Pacific Journal of Marketing and Logistics, 34*(5), 865–886. https://doi.org/10.1108/APJML-01-2021-0016.

Watson, R., Wilson, H. N., Smart, P., & Macdonald, E. K. (2017). Harnessing difference: A capability-based framework for stakeholder engagement in environmental innovation. *Journal of Product Innovation Management, 35*(2), 254–279. https://doi.org/10.1111/jpim.12394.

White, K., Habib, R., & Hardisty, D. J. (2019). How to SHIFT consumer behaviors to be more sustainable: A literature review and guiding framework. *Journal of Destination Marketing and Management, 83*(3), 22–49. https://doi.org/10.1177/0022242919825649.

3 Differences That Make a Difference

Placemaking Through Imaging and Imagination

Marc Boumeester

Introduction

Tourism and art both deal with expectation, albeit in an almost antagonist manner. Travel for leisure is oftentimes fueled by the excitement of encountering a well-defined set of effects which often have been heralded by an abundant "imaging" that portrays the destination. Matching the existing with the expectation thereof (the "if this, then that" epitome) is therefore of the highest importance to fulfil satisfaction. In contrast, and despite popular belief, art is not intended to reproduce the already existing, but to quote Swiss-German painter Paul Klee (1920, p. 28.): "Art does not reproduce the visible; rather, it makes visible."[1] It is for this reason that a long-lasting tradition in the arts could emerge whereby the concept of travel was interpreted as imagination rendered solid, an allegoric embodiment of non-experienced experience, that allowed writers and visual artists to explore, interpret and imagine the world in one single movement (the "what if…?" paradigm). Some of its better-known forms are *psychogeography* and *armchair travel*, whereby the first reinterprets the existing urban fabric on the basis of effects, rather than on actualized facts, and the latter refers to the creation of travel-logs and other depictions of travels that were never physically undertaken. *Armchair travel* often leads the artist to a point beyond the state of current knowledge or understanding, with exceptionally imaginative and futuristic examples in the eighteenth and nineteenth centuries, in which this type of travelling could easily lead into "parallel worlds".[2] Both expressions serve as responses to an overly rationalized worldview and offer alternatives through metaphorical embodiments of abstracted effects. Alongside the opportunities to travel, which have been steadily growing in the twentieth century, armchair travel has changed its character. Artists like Joseph Cornell and Rinus van de Velde embraced this modus of travelling in their praxis on conceptual grounds, claiming that the original experience can only diminish the beautiful images that have been made of it. Imaging instead of actualization is the highest form of respect. Radical perhaps, but this notion functions as a waypoint to the concept of *intensive* thinking as an alternative to the *extensive* thinking that has dominated many worldviews for centuries and which lies underneath many of the current problems that threaten our societies and lifeforms.

In short, extensive thinking classifies a world on the basis of properties and attaches values to them following topographical patterns that deal with representations

DOI: 10.4324/9781003449027-4

and imagings of (physical) features and their (social, cultural, economic) significations in discrete and actualized domains. Conversely, the intensive is immersive and
imaginative, and it is characterized by its topological and transformative descriptions
which focus on the underlying relationships between elements and their capacities,
rather than on any of their properties or significations. Extensive thinking creates
causal premises which will produce a difference (*if* I travel to Berlin, *then* I will finally
see the Brandenburger Tor), whereas intensive thinking will bring inquisitive premises (*what* will happen *if* I travel to Berlin?) which produce "differences which make
a difference" to paraphrase Bateson (1972, p. 272). In the first case, the degree of
expectation fulfilment (which is extensive and topographical) will determine the level
of satisfaction, and in the latter the level of *un-expectation* (which is intensive and
topological) will determine the mark of gratification. Note that the concept of intensive thinking as used here should not be confused with intensive tourism, but rather
the opposite: the core of this argument is that spending a week in a radius of 100 meters around a hotel will teach one more about the culture and people of Berlin, than
spending the same time chasing all the "highlights" that this city has to offer.[3] This
chapter hopes to inspire a transformation of the premises of travel for leisure from the
inevitable "if then" into the more elusive "what if?" by introducing a few concepts.

What if we were to search for an intensive and totipotent gratification in tourism with a future-based intensive outlook rather than with a traditional extensive perspective?

The quality of the question determines the quality of the answer, and if anything, it
is the task of the artist to formulate better questions than more appropriate answers.
The main thread in this chapter is therefore the question; "What if?" and it will
expand on three sets of interconnecting concepts which can be helpful in defining
better questions that can lead to improved solutions for the existing issues that burden both the tourism industry and the territory it operates in. To match the urgency
of the challenges faced by our planet with the speed of our problem-solving efforts,
a symbiotic bond between theory and creation must be forged. This bond calls for
the development and conceptualization of new educational approaches in this field,
emphasizing learning through practical *creation* rather than through reproductive
learning. Artistic research is such a praxis and merges creative experimentation,
critical reflection and aesthetic exploration with academic research methods to
generate new knowledge and understanding in broader dialogues with academic,
cultural and societal contexts. Drawing from this, some praxes in artistic research
will be brought forward to exemplify the concepts that are introduced.

Perception and Expectation

At the core of my arguments lies the belief that both the virtual (potential) and the actual
are perceived as realities, with no existence beyond this realm. The actualized is tangible and concrete, encompassing energy, matter, movement or abstract concepts – it
undeniably exists. On the other hand, the virtual "subsists"; it is an intensive and
imaginative agency, operating in the realm of possibilities, or as philosopher Karen

Barad (2013, p. 52) summarizes: "Agency is not held, it is not a property of persons or things; rather, agency is an enactment, a matter of possibilities for reconfiguring entanglements". This approach demands a view in which everything real is contingently obligatory and discards the theorem of logical necessity as this type of logic (if, then) falls short due to its oversimplified understanding of the intricate interplay between forces, drives, agencies and conflicts that shape the intricate tapestry of life. By recognizing the fluidity and transformative nature of life, we can delve deeper into understanding its complexities; instead of fixating on the extensive environment itself, the focus should shift towards perceiving life as an intensive, dynamic, creative process in a constantly evolving assemblage of stimuli.

The movement in any journey comes with an unlimited array of stimuli. Any number of vantage points, perspectives, sounds, smells, tastes, etc., will enter the realm of our somaesthetic perception.[4] However, these are only the elements that are actualized, and apart from them we have a far greater quantity of pre-/post-stimuli that lie in the domain of the virtual. These stimuli emanate from past and future experiences (memory and expectation) and it is needless to say that these stimuli are highly individual. It is for this reason that we can safely claim that no experience is ever the same, not for different participants of the same event, nor for the individual itself. An ongoing and intensive process gives rise to coagulations of actualizations and a continuous movement of inclusion (autopoiesis) and exclusion (entropy) of what the human sensorium will perceive. Through this process, an intriguing paradox arises: a "limitless limitation" emerges as an unlimited array of new dimensions accumulates yet yields only a finite number of outcomes. This is a dynamic process in which all stimuli, whether actualized or virtual, are combined, recombined, interpreted and reinterpreted into singular cognition. For this reason, we need to emphasize the distinction between perception and cognition, where perception is the process of receiving and sifting somaesthetic sensory information, and cognition refers to the broader range of mental processes involved in acquiring, processing, signifying and utilizing information beyond sensory perception.

Imagine a waiting area at Dresden train station where voyagers are waiting for the 11:10 EuroCity 173 international train to Prague. All are seemingly exposed to the same stimuli: the chattering of voices, the smells coming from the coffee stand, the occasional announcement through the intercom, the squeaking of the brakes of arriving trains, the view of other travellers, the temperature and humidity, and so on. Probably most are somehow occupied, on their phones, in conversation, in eating a sandwich, reading, glancing around or listening to music. Simultaneously all are affected by elements that come from the past: the unsettling feeling after saying goodbye, the relief that we arrived at the station on time, did I lock the house twice?, etc. Besides that, anticipations of future events are also at play: the joy of freedom this journey will bring, the excitement to see an old friend, the curiosity of if the hotel will live up to its reviews or the somber prospect of potentially having to sit next to a particularly unfriendly looking fellow traveller. Amidst all these actualized and virtual stimuli, random thoughts and emotions bump into the mind: a seemingly forgotten song comes into memory, did I fill out my tax report right?, purple is a nice color, etc. Thus, this seemingly identical shared experience is already fully individually perceived, and this isn't even taking into account how all these stimuli are mentally processed by storing, signifying and utilizing this information. All the infinite possibilities that could shape the cognition of this event will – without exception – come to one finite individuated version of the event. The illogic of the theorem of logical necessity (if then) is shown here, as it is simply not enough to consider all the givens (the conditions, the territory, the destination) to determine the outcome of an experience: the event is always bigger than the sum of its parts.

What if we were to encounter a touristic environment with an approach that tries to grasp and enhance the surplus that forms the intangible part of the event, rather than to focus on the formal properties of that event?
To both unravel and enhance this question, I have started many projects with students involving a new territory by conducting the "six-minute workshop". This entails the physical exploration of the site and concentrating on one of the senses at a time, each for the duration of one minute. The impressions are noted on a piece of paper and collectively discussed afterwards. The sixth sense is also (literally) considered, hence the six minutes. The sixth sense colloquially indicates the sensitivity to perception caused by other senses than the ability to see, hear, touch, smell or taste. Also known as extrasensory perception, it is often associated with premonitions, intuitions, paranormal phenomena and the occult in general. In this case it indicates the "feeling" one gets in a certain place, the sensation that acts before sensing consciously. The performance of exploring an area by licking, touching and smelling obviously drew quite some attention in places like Hong Kong, New York, Stuttgart, Nanjing and others, but the experience brought many new insights, which – most importantly – dislodged some of the existing assumptions of the participants. The somaesthetic perception gave in many cases completely different information than was anticipated, which proved vital in the following design process as assumptions subconsciously already pre-formatted a solution, whilst the empirical data posed different questions. In the artistic research program "Ecologies of Performance" for instance, students of various institutions collaborated with the municipality of Florence to investigate the issue of "hit-and-run" tourism.[5]

In this hands-on artistic research program, the groups collaborated directly with the touristic division of the municipality. By exploring the performance ecology in Florence – as a tourism destination – the goal was to understand the interactions between performances, visitors and resident communities based on empirical somaesthetic input rather than on abstract properties and (historical) significations. The project mapped how different community economies coexist, challenge or threaten each other, prompting reflection on inclusive practices. The results of this exercise in transversal questioning were feedbacked to the municipality. Considering the ways in which perception, memory and anticipation are transformed into cognition is a valuable element in formulating the questions of tourism adaptation and transformation. The notion of *intensive* as an alternative to *extensive* requires different ways of expectation management that can positively influence the problem-solving process. The role of imaging in the creation of identity is crucial in the process of building expectations and will be addressed in the next section.

Figure 3.1 Ecologies of Performance, 2019

Source: Photo by Fenia Kotsopoulou with permission by Home of Performance Practices

Figure 3.2 Ecologies of Performance, 2019

Source: Photo by Fenia Kotsopoulou with permission by Home of Performance Practices

Figure 3.3 Ecologies of Performance, 2019

Source: Photo by Fenia Kotsopoulou with permission by Home of Performance Practices

Figure 3.4 Ecologies of Performance, 2019

Source: Photo by Fenia Kotsopoulou with permission by Home of Performance Practices

Figure 3.5 Ecologies of Performance, 2019
Source: Photo by Fenia Kotsopoulou with permission by Home of Performance Practices

Imaging, Image by Proxy and Exo-Identity

The use of (snapshot) photography to capture experiences is not a new phenomenon. Postcards, holiday slide shows for the homestayers and holiday-photo reunions have long served as imaging instruments. Historian-philosopher Verena Winiwarter (2000) has pointed out that such imaging is never intended to be a neutral representation but rather acts as a mediator between tourist expectations and the reality of the vacation spot. What sets the current era apart is the pervasive ability of individuals to mediate between their expectations and experiences. The history of photography of roughly 100 years has generated approximately 12.4 trillion photos, of which roughly 72% were produced in the last decade alone – a gargantuan acceleration of production that has little comparison. This number is projected to increase up to 2 trillion annually, even without taking image generation by artificial intelligence into account. The immense volume of images available on the internet, including those shared on platforms like Instagram, TikTok and WhatsApp, highlights the unprecedented scale of this phenomenon. Therefore, it becomes crucial to set these "individual" images within the context of the vast collective of imaging and acknowledge its significance within the Anthropocene epoch. To that end, we can detect a relatively new type of image in the taxonomy of images which is called the "image by proxy" (Boumeester, 2023). The term "by proxy" denotes the system of the production of individual images as (un)intentional replicas of existing images, imbuing them with a sense of proximate significance granted by the vast volume of preceding images and their imaging agency.

Imaging is an umbrella term for the formation of (mental) images of a future event and is thus key in building the expectation thereof. Imaging contains the acts of seeing images, hearing stories, reading news or reviews, experiencing some elements of the culture of a place, etc., that all lead to building the expectation of the actual encounter of that destination. When visiting tourist destinations for instance, expectations are often shaped by images depicting those places, and to validate their experiences, visitors often capture their individual albeit very similar photographic snapshots. Although these images portray an individual experience ("me in front of the Eiffel Tower"), they are a repetition of a close to endless row of images of people in front of the Eiffel Tower. Their academic significance thus lies less in the individuated version of "the Eiffel Tower photo" and more in their place in the system of producing photos of the Eiffel Tower on this scale. The significance is thus given "by proxy" by the large volume of similar photos – the image itself is an "any-image" of that event. The collection of these individual images contributes to the imaging and formation of identities that influence the expectations of other visitors, especially when shared on social media. On a larger scale, this process of identity creation impacts population mobility by the formation of worldviews and other information biases. This has become painfully clear with the advent of images produced by artificial intelligence which, without exception, draw from the aforementioned large volume of available digital/digitized photos as a source of "learning" in how to construct new images. As a result, images are produced that affirm and propagate representational biases, creating "representations" that are heavily gendered, racially and ethnographically biased, and obviously these images themselves become part of the new "learning curve" of the next generations of images, so that stereotypical "identities" will only be reaffirmed repeatedly. As argued, the blueprint for "the image by proxy" already existed before the digital age, but its force and speed have been rapidly increased by the current information architecture, which will only fuel itself to even greater speeds and larger volumes, which is far from harmless.

The recursive interplay of imaging, expectation and perception shapes an individual's reality both mentally (image-expectancy) and physically (image-perception). Moreover, the image by proxy influences spatial awareness by pre-defining angles, frames, scales and perspectives: a tourist destination that has been extensively documented through imaging gives little to no opportunity for an unbiased experience. According to theorist and architect Rem Koolhaas, robust urban identities possess the ability to withstand changes such as growth and rejuvenation. Koolhaas (1998, p. 1239) suggests that "these identities can be likened to an overpowering lighthouse, exerting a significant influence on their surroundings … it can change its position or the pattern it emits only at the cost of destabilising navigation". Koolhaas (1998, p. 1240) continues "Paris can only become more Parisian. It is already on its way to becoming hyper-Paris, a polished caricature. There are exceptions: London – its only identity a lack of clear identity, is perpetually becoming even less London, more open, less static." The caricature Koolhaas is sketching refers to what is known as an exo-identity (Boumeester, 2022), which is a specific way of placemaking by imaging. This is an identity that is based on the expectation of that identity, rather than on the evidential identity. An exo-identity is a product of extensive thinking and beholds a topographical description of properties that are supposed to make a place. It is an imaging of culture, language, habitation, behavior, social conditions, etc. that only exists on an abstract and often mediated level, causing a biased perception of a particular place. Exo-identities can exist for decades or even centuries: "*Berlin ist eine Stadt, verdammt dazu, immerfort zu werden und niemals zu sein*" (Scheffler, 1910/2015, p. 219),[6] and "Paris is always Paris and Berlin is never Berlin!" (Lang, 2001)[7] are just two of the countless quotes that affirm an exo-identity and are in that sense part of the proximation system of the imaging of a city. The exo-identity is created and maintained by several stakeholders concerned with imaging the specific place: City marketeers highlight the already known features, tourist operators highlight the already known experiences, hoteliers highlight their proximity and connection to known landmarks, etc. In this way, the identity of a place is enforced, repeated and secured time and time again, leaving little room for the unknown. The performance of a ritual dance in Bali for the benefit of tourists might be lacking taste, yet it remains somehow attached to the inherent culture of the area and feeds into the local economy. Selling miniature towers of Pisa as souvenirs might be considered tacky, yet still it is grounded in the event of the actual visit to the original (very few travellers will visit Pisa and come back with a miniature Eiffel Tower). There can thus be a gradual shift from identity to exo-identity and if it is "authentic" it will – at least partially – benefit the local community, as tourism has in many cases already been a part of local cultures for many decades.

It becomes far more dubious when these systems have not grown "naturally" but are constructed for self-beneficial purposes only. The *International Style* in architecture deliberately dislocated the identity of the construction from its historical and cultural setting and it can be said this method can also be found in the construction of consumption and tourism in the last decades. Large chains of (fast food) restaurants, hotels, bars, coffeehouses, museums, retail stores, Instagram décor-centres, floating "entertainment palaces", and other industrial consumer precincts that Marc Augé (1992) would name "non-places", have encompassed the globe as their territory, turning any city into "any-city": a dislocated, unrooted and unauthentic event machine that produces sensations with the highest possible saturation of stimulation. This is highly effective, but by nature dominant and possessive, leaving no room for the development of true local interaction, which profoundly affects both inhabitants and visitors and their right to act, look and feel differently.

Tourists also reinforce exo-identities by their extensive search for highlights and the active imaging of themselves doing so. Tourists are often looking for confirmation of their expectations and in doing so, help to strengthen those expectations for others. As argued, it seems that no experience is complete without a picture with La Tour Eiffel, the Fernsehturm or the Karlův Most in the background. These images (by proxy of an exo-identity) are made individually, yet the drive to produce them is to approximate the identification of being part of the event. The "if then" dogma works on both sides – the interlocking assembly of stakeholders reinforce the exo-identity of a place, creating an experiential bias and fulfilment system. This is where the collision with the residential stakeholders starts: the inhabitants, workers, shop owners, students, bus drivers and all others who rely on their city to be flexible and in motion. All those who have a need for progression and development in one way or another, those whose movements are characterized by the *non-highlighted* events that form daily life. For them a role in a flattened exo-identity is unhelpful, insulting even, let alone for those who have a completely different cultural reference and see their holy places overflooded and rituals exploited by visitors who have no inherent understanding of or respect for their setting. As artificial as the smile on a selfie mostly is, just as fake is the 9 billionth image of a cappuccino on an allegedly undiscovered terrace (extensive) as an alternative to the actual experience of presence (intensive).

The concept of presence emphasizes the unembellished importance of rootedness, highlighting the impact of humanity on its territory and its potential to shape alternative futures. Presence exists solely through action and relationships with others and the territory, focusing on the level of engagement and investment from both creators and audiences, which diverges from dualistic cultural theories by recognizing the agency (of matter) itself. It shifts the focus from anthropocentric interpretations to a shared realm of non-signification that encompasses matter, medium, mind, and body. This narrative goes beyond biology, natural ecology and sociology, encompassing empowerment, aesthetics and locality in search of a tangible and non-commodifiable presence which leads to what Isabelle Stengers (2018, p. 143) names honoring of divergences because" what can enter into communication with the word "honor" is something that will be apprehended *not as a particularity of the other, but as what the other makes matter.*" This outlook empowers an ongoing process of reconfiguring entanglements, allowing for an unbiased and inclusive understanding of the present moment rather than an illusory proof of concept. It is a shift from imaging to imagination and from fulfilment to curiosity.

Figure 3.6 Ecologies of Performance, 2019

Source: Photo by Fenia Kotsopoulou with permission by Home of Performance Practices

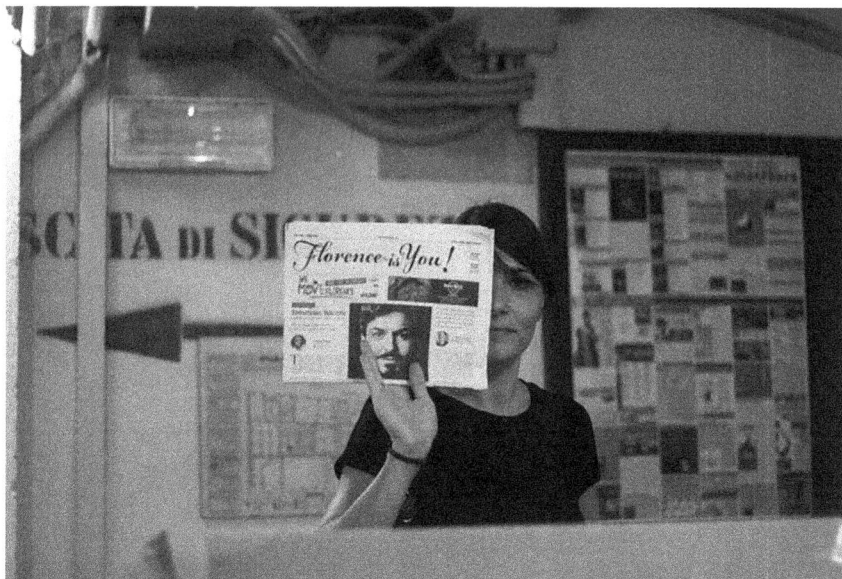

Figure 3.7 Ecologies of Performance, 2019

Source: Photo by Daz Disley with permission by Home of Performance Practices

What if we were to centralize the notion of presence in the building of experience, acknowledging the imaging (by proxy) in the making of expectation, yet disconnecting this system from the actual anticipation?

In an attempt to disentangle this question, an artistic research project started in 2022 at the Het HEM museum in Zaandam (The Netherlands) in collaboration with ArtEZ University of the Arts. The project aimed to explore the disparity between mediated expectation and real-life experience (presence) itself. Drawing inspiration from the research exhibition "Chapter 5IVE" by Samir Bantal and Rem Koolhaas, which delved into the idea of complete environmental control, the participants focused on critical themes such as nuclear waste storage, common language issues, non-human perspectives on industrial site redevelopment, the impact of technology on our perception of the environment, and how dreaming can redefine healing for a collective future. These are no trivial topics and reflect the state of urgency that the researchers experienced before and during the process. When the severity of the research topics was measured against the public yet relatively safe space of the museum, the presence of the place became a vital design instrument. By working in the actual exhibition space, the participants were stimulated to fully absorb its presence and use the confines of the "white cube" to seek shelter against the mental impact of their findings. Working amongst visitors and staff, the place became its own source of input as they were invited to reflect on the work in progress, altering the expectations of all parties involved. Yet, set against both the significance of the place itself and the status of both curators and the museum, the experience could not be free of contextual anticipation. Therefore, a year later, in 2023, the same group gathered in a different location, a communal centre out of the limelight run by Pakhuis de Zwijger in a western area of Amsterdam called "Nieuw-West." This multicultural area faces socio-economic disparities but is actively engaged in community initiatives and events. Immersed in the site, the researchers adopted a six-minute workshop approach to challenge their preconceptions and initial anticipations. Perhaps unsurprisingly, the site defied expectations (its exo-identity), yielding positive effects and percepts, offering a haven that transformed various flows and drives. Throughout their research, the participants built a cartography

Figure 3.8 The project at the Het HEM museum, 2022
Source: Photo by Polina Nasedkina

Figure 3.9 The project at the Het HEM museum, 2022
Source: Photo by Polina Nasedkina

Figure 3.10 The project at the Het HEM museum, 2022
Source: Photo by Polina Nasedkina

of the residents' and businesses' mental and physical flows, involving the community in the process. The collaboration led to a reciprocal debunking of expectations, with the artistic researchers and the community learning from each other. The project in Florence, at the Het HEM museum and the subsequent exploration in Nieuw-West demonstrated the sway of artistic research as a mode of understanding the interplay between anticipation and experience in a socio-economic context.[8]

Totipotence, premeditated obsolescence and the desire of plenitude

The last three concepts that are introduced in this chapter attempt to enclose and define a field that lies at the center of our problem. These notions stem from different disciplines in the belief that any new answers cannot be found in the existing extensive approach (including its disciplines) but must lie in the external (intensive) relations that can only be found *in between* the elements at play. Therefore, we start with the questions and track them with the concepts.

What if, instead of constantly inventing something new, we enhance the existing in a systemic and sustainable way, thus creating a desire of plenitude, rather than a desire of lack? What if we could learn from the notion of planned obsolescence as a form totipotence to achieve this?

When in 1924 the American automobile market began to reach a point of saturation, General Motors executive Alfred P. Sloan Jr., proposed to introduce the concept of "dynamic obsolescence". In short, the concept proposed to renew car models annually, albeit only in a superficial esthetical way. This way the consumer was lured into buying a new product much sooner than technically needed – just to keep up with the flavor of the time – and indicated the financial capacity of the owner, therefore appealing to their vanity. Critics later popularized the term "planned obsolescence" to describe his strategy – focusing on its aim to endorse overconsumption and create overproduction – and this system is largely maintained in many industries to date. Despite its heavily tainted connotation and devastating effects, it would be a missed opportunity to categorically discard the process of premeditated obsolescence without having examined its key components, as desire is at the heart of that process.

Desire is an existential state that strives for its own independence and emergence, without a fixed starting point or a predetermined endpoint. Desire thus relies on its inability to be fully satisfied. Current ways of consuming create what Jacques Lacan (1980/2006) calls a "desire of lack", meaning that the desire is by default produced beyond the point of its fulfilment and thus basically acts like any other addiction.[9] This mode of consumption creation has had detrimental effects on – literally – the world; it needs no argument that the current global challenges (climate collapse, biodiversity breakdown, clean water shortage, social injustice, postcolonial inequity) have many of their roots in the inflamed drive for progression in capitalist terms. Endeavoring to "fill the lack" is the key selling capacity of conducting business, and therefore this lack is continuously artificially created. To a certain extent this system applies also to the tourism industry, with the mediated system of imaging creating a (social) desire of lack. As a challenge to the Lacanian view of the desire of lack, psychoanalyst Félix Guattari (2009) describes desire as encompassing everything that exists prior to the division between subject and object, before representation and production come into play. When an event is synthesized, it generates an excess (of stimuli) that surpasses what can be predicted or measured based on the sum of its individual elements. In an extensive view of an event, this surplus would be disregarded as it is not part of the fulfilment of expectation, yet with this action, valuable commodities are being wasted. This is called a desire of plenitude and is continually generated, therefore it has no pre-defined dimensions.

The last of these three concepts is called "totipotence", a term stemming from biology and introduced into philosophy by Daniel W. Smith (2018), which denotes the capability, within the realm of the actual, to embody diverse entities. A totipotent cell cannot only produce any desired specific cells by dividing itself into that, it can also produce itself, and is therefore its own genie in a bottle. It does not involve the realization of pure potential, as both the preceding and succeeding states have already materialized in some form. Rather, it entails the transformation of already actualized matter into an alternative ontological framework, wherein the nonspecific experience or effect takes shape. To paraphrase Stengers (2015, p. 143) once again, it honors divergence as *what the other makes matter*, not as a particularity.

Synthesis

To shift from an extensive experience to an intensive experience requires that the event itself must remain more and less the same, as changing the event would create a new lack. The nuance lies in the more *and* less because the variation resides in the perception, not in the event itself. The aim would be to create a desire for plenitude (to honor divergence) which would feed into the progression of the experience of the consumer. It is essential, of course, that the previous experience must contain a planned obsolescence and is no longer desired, with every next step from the last experience left hungry for the same albeit different experience. Been there, and done that, yet will go back to do the same but differently.

An example of a desire of plenitude in tourism I encountered some years ago, was when I queued up behind a family travelling from the United Arab Emirates, with a destination in Scotland. Their choice for this was determined by the probability of rainfall during their stay, a sensation they normally hardly ever experienced. The fact that they would be staying in a breathtaking landscape and could indulge in the local "*boisine*" was a bonus, not a goal. They had been visiting the same place several times before and gradually their experience had gained more depth as well as their understanding and respect for the local culture. This was only possible because the initial drive to visit this location was to enable the sensation of being in the rain – an intensive experience par excellence as it is non-specific and non-local, yet the locality enables this particular event. The growing attachment to the "subsidiary" conditions made the totipotent event modulate and produce new outcomes time and time again. For them, there was no need for another destination, as this one provided different experiences that grew in value over time, making every previous experience obsolete yet leaving a desire for the next. The event itself was totipotent – producing different experiences time and time again, which are in themselves planned obsolescent.

Conclusion

Tourism has in a sense become an exo-identity itself; the global sameness and unrooted motives for travel are often fueled by imaging that is causing expectation and perception to become one, reinforcing stereotypes to the point of hollow and trivial imagery (any-photo in any-city). Challenging as it may be, a new paradigm is needed that sheds ties with old patterns of extensive thinking. The real problem lies not in the satisfaction of needs, as these are driven by a desire of lack and are therefore unsolvable. The abilities to create differences as such are amply available, yet differences that make a difference are still rare. Besides that, geopolitical tensions and the awareness of global interdependence, both ecological and economical, are unmistakably foreshadowing a need for an overhaul of the ways needs for leisure can be satisfied. As seen during the last large pandemic, all "givens" can become antediluvian in an instant and the effects of recent war have overnight changed the landscape of travel destinations for many years to come. Yet the greatest challenge of all is the swiftly changing outlook on how we have been treating

(a)

(b)

(c)

(d)

Figure 3.11 Totipotence in Nieuw-West Amsterdam, 2023

this planet, its consequences, and the ostensible resistance of many large systems to act on this in a sincere and adequate way. The map of destinations, their qualities and their (exo-)identities will be changing more and more because of changing climates, which is by now no longer a daunting foresight, but a hard reality.

I have highlighted several concepts that are helpful in formulating new questions regarding the challenges of the tourism industry. The interventions described here clearly demonstrate the strength and applicability of artistic research in this process. Although these interventions still operate on a small scale, they are not insignificant: as the process of placemaking is continuous and relentless, it also gives the opportunity to renew "from within", which can have an asymmetrically large effect. Acknowledging the necessity to replace the hunt for the extensive experience with a far more satisfying and sustainable intensive perception is a crucial step to reformulate goals and objectives. As argued, no event, no image, no spectator or experience is ever identical, and under a transition from *imaging* (if then) to *imagination* (what if?) experiences with a rooted presence still offer a vast potential for development and expansion; the future of tourism is in the differences that make a difference.

Notes

1 Original in German: "*Kunst gibt nicht das Sichtbare wieder, sondern macht sichtbar*".
2 Étienne-Gaspard Robert, Jules Verne, Arthur Machen and several others explored the realms of "parallel worlds" which (in different interpretations) create an extension to the

possible. Genres such as science fiction, occultism and horror were established through their writings or performances.

3 Intensive tourism is a colloquial container term for tourism based on short, invasive, mono-cultural and superficial "use" of a tourist destination. Well-known examples are cruise ships that unload a bulk of passengers for a very limited amount of time to a pre-arranged set of shopping opportunities (hit-and-run tourism), or the phenomenon of TikTok tourism that draws large numbers of visitors to one single, rather mundane destination – like a coffee bar – with the sole purpose of acquiring some items that can be displayed on a TikTok or Instagram feed.

4 Somaesthetics is an interdisciplinary field of study that focuses on the relationship between the body (soma) and aesthetic experiences. It explores how bodily perceptions, movements and sensory experiences contribute to our understanding and appreciation of beauty and aesthetics in general.

5 This program was orchestrated by ArtEZ University of the Arts, the Master "Performance Practices", Buas and Teatro della Pergola in Florence, Italy. https://ppf.artez.nl/ecologies-of-performance.

6 Translation: "Berlin is a city doomed eternally to become, yet never to be".

7 Quotation from a speech Jack Lang gave as Minister of Culture in France.

8 The projects in Florence, Het HEM and Pakhuis de Zwijger are initiated and developed by Carin Rustema as an initial step to address socio-economic challenges by inserting artistic research directly in the domain of the partners. It expands on McLuhan's *City as Classroom* and a holistic no-school principle and is characterized by exterior learning, e.g., what to learn is generated by conditions, how to learn is generated by capacities, validation is done on basis of growth, not on achievement. https://hethem.nl/en/Chapter-Five/Summer-Research-Programme; https://dezwijger.nl/update/open-for-application-community-research-programme-artez.

9 In Écrits, chapter 24, "The Signification of the Phallus", pp. 575–585, Lacan (1980/2006) remarks that "desire is neither the appetite for satisfaction nor the demand for love, but the difference that results from the subtraction of the first from the second".

References

Augé, M. (1992). *Non-Places: Introduction to an Anthropology of Supermodernity.* Verso.

Barad, K. (2013). Matter Feels, Converses, Suffers, Desires, Yearns and Remembers. In: *New Materialism: Interviews and Cartographies.* R. Dolphijn & I. van der Tuin (Eds.). Open Humanities Press. pp. 48–70.

Bateson, G. (1972). *Steps to an Ecology of Mind: Collected Essays in Anthropology, Psychiatry, Evolution and Epistemology.* The University of Chicago Press.

Boumeester, M. (2022). Cinematographing Perception: Autopoiesis and Entropy of a Mediumless Medium. In: *Proceedings "On Cinema Conference"* at Centro de Estudos Arnaldo Araújo, Escola Superior Artística do Porto, Portugal, pp.42–56

Boumeester, M. (2023). Technicity as the Montage Production of the Mundane. In: *The Space of Technicity: Theorising Social, Technical and Environmental Entanglements.* R.A. Gorny, S. Kousoulas, D. Perera & A. Radman (Eds.). Rowman & Littlefield.

Grattan, L.G. (2016). *Populism's Power: Radical Grassroots Democracy in America.* Oxford University Press.

Guattari, F. (2009). A Liberation of Desire. In: *Soft Subversions: Texts and Interviews 1977–1985.* S. Lotringer (Ed.). Semiotext(e), pp. 142–145

Klee, P. (1920). Schöpferische Konfession. In: *Tribüne der Kunst und Zeit.* K. Edschmid, Berlin (Ed.). Erich Reiss Verlag, pp.28–40.

Koolhaas, R. (1998). *The Generic City: SMLXL.* Monacelli Press.

Lacan, J. (1980/2006). *Écrits*. W.W. Norton.

Smith, D.W. (2018). Deleuze, Technology, and Thought. *Tamkang Review* 49(1), 33–52.

Stengers, I. (2015). *In Catastrophic Times: Resisting the Coming Barbarism.* Open Humanities Press.

Winiwarter, V. (2000). Landschaft auf Vierfarbkarton. *Zolltexte* 10(35), 48–52.

Scheffler, K. (1910/2015). *Berlin. Ein Stadtschicksal*. Suhrkamp Verlag.

4 Intervention(s) for Island Community Survival Through Tourism

The Case of the Fair Isle Bird Observatory

Richard R. Butler

Introduction and Context

In this chapter the specific intervention being discussed fits several of the definitions cited earlier in this volume, the most appropriate one probably being that of Midgley (2000: 113) who defined an intervention as a "purposeful action by an agent to create change". In this case, the intervention was both purposeful and designed to create change, involving the creation of an establishment with a goal of ensuring the stability and viability of an island population, in part through developing tourism. It also fits the expanded definition of Moretti (2021: 5) in that it was a "purposeful action ... proved to contribute (and) was designed to contribute ... (and) is still ongoing) to the socio-cultural landscape".

The original intervention (which has seen continual modification and reconstruction over the last eight decades) was the creation of one individual, George Waterston, and was supported, after he relinquished ownership and responsibility for the island, by two charitable organisations, the Fair Isle Bird Observatory (FIBO) Trust and the National Trust for Scotland. Waterston was a leading figure in Scottish ornithology, founding several related organisations and became the first full-time Scottish representative of the Royal Society for the Protection of Birds. He was responsible for the successful return of ospreys to Britain in the 1950s and is recognised on Fair Isle by the Memorial Museum bearing his name. Other aspects of his life are recounted in Niemann's (2012) book *Birds in a Cage* (and in FIBO annual reports and materials in the FIBO archives), which deals primarily with the period during the Second World War when Waterston and several other soldiers interested in birds were prisoners of war in Germany. During that time Waterston developed the idea of purchasing Fair Isle and establishing a bird observatory there, comparable to the then famous one on the island of Heligoland. When he was released because of ill-health in 1943, on his way back to Britain, his first sight of the British Isles was of Fair Isle and the idea was rekindled.

As will be discussed, Waterston was clear in his intent when making the intervention and pursued his goal through an approach that best is described as *sustainable development* long before that term was coined, in the sense that it incorporated

DOI: 10.4324/9781003449027-5

a long-term viewpoint, integrated objectives, and local stakeholder participation and operation within natural and cultural limits.

The island in question, Fair Isle, is the most remote of the inhabited of the British Isles, lying midway between the Scottish island groups of Shetland to the north and Orkney to the south. It is located at 59 degrees north, some 40 kilometres from its nearest neighbour (the Shetland "mainland") and is approximately 5 kilometres in length and 2 kilometres in width, with a current population of around 65. The island is more widely known than might be expected through the long-standing popularity of Fair Isle knitwear and patterns, and because it has also given its name to one of the designated weather regions used daily by the BBC in its shipping forecasts. To birdwatchers it is famous for having added more new species to the British list of birds than any other area of comparative size, reflecting its location on a major bird migration flyway and is thus visited frequently by those interested in ornithology, a growing form of niche tourism (Novelli and Benson 2005).

Fair Isle, before the intervention discussed here, like many small islands with limited resources, faced difficulties in surviving in the modern world (Alberts and Baldachinno 2017; Amoamo 2017). Its economy had traditionally been dependent on crofting (small-scale tenant farming) and fishing. The population had also gained a small amount of income from handicraft production, particularly knitted goods that were traditionally traded with passing ships until the 20th century when this ceased, as few ships then called at the island. Ironically, this situation has now been reversed with cruise ships providing regular visitors and purchasers of the knitwear especially, as noted below. A population of over 300 in the early 19th century had seen a major emigration of 130 people to New Brunswick in Canada in the 1850s (Willis 1967) and had further declined by the immediate post-war period to below 40. The island had been owned by a family on the Shetland mainland for many years with little having been done to improve the lot of the tenants through land or building improvements, and restrictions had been imposed on trade with vessels. The threat of evacuation of the island, as had happened to St Kilda to the west in 1930 when its population fell to 36, was a strong possibility, and one of the reasons for the intervention which took place.

While the Second World War saw a temporary boost to the population of the island by virtue of military forces being garrisoned there, the economic outlook in 1945 was poor. The knitting industry was plagued by very low prices paid to knitters by agents in London, with their products being sold for grossly inflated prices to major retail outlets. Thus, the local producers gained little from the rise in popularity of Fair Isle knitwear, boosted by the wearing of such products on public occasions by the then Prince of Wales, the later King Edward VIII (Thom 1989). Crofting produced small returns for tenant farmers, mostly from sheep and wool, and much of the rest of the diet came from local fishing and subsistence level vegetable production. In the 1940s there was no electricity supply across the island, generators provided power to the two lighthouses with some of that power being made available at limited times to island houses, and most goods and fuel had to be imported at considerable expense.

There was no tourism to the island before 1948 and very few visitors, almost all of whom were either professional or amateur scientists, and a few writers such as Sir Walter Scott who visited the island in 1814. A number of visitors interested in birds came to the island during the first quarter of the 20th century (Scott and Thigpen 2003). Eagle Clarke, a distinguished naturalist, was the first and his records indicated the considerable scale of bird migration experienced and the numbers of species involved. He made several visits to Fair Isle between 1905 and 1911, recording a remarkable 207 species (the current total stands at 345), representing half the total British list of species at that time, and arranged for one of the islanders to trap or shoot other birds for him to identify and record on later visits. His interest was shared by the then Duchess of Bedford who made nine visits to the island between 1909 and 1914 by private yacht, staying at Pund (see Figure 4.1), an empty croft (Duncan 1938). Following the end of the First World War, the island was visited several times by Rear Admiral Stenhouse between 1921 and 1927, and it was he who introduced George Waterston to the island, thus providing the catalyst for intervention. Waterston had met Stenhouse as a schoolboy and first visited the island in 1935, stimulating an interest and commitment to the island for the rest of his life.

Fair Isle Bird Observatory: First Intervention

Waterston was a central figure in the development and survival of the Fair Isle community until his death in 1980. He purchased the entire island in 1948, including what were formerly naval huts in the north part of the island (Figure 4.2). These were converted into the first Observatory, and this was opened on 28 August 1948, being operated by a charitable trust established by Waterston. The Fair Isle Bird Observatory Trust remains the responsible organisation and is run today by a Board of Trustees (Butler 2014). Waterston's intention in buying the island was clear, although his original plan had been for the Observatory to be located at Pund, the croft used by the Duchess of Bedford. He discovered that during the war the army had burnt down the croft and it was unusable (Niemann 2012: 149), and the former naval huts at North Haven (Figure 4.1), perhaps fortunately, for reasons discussed below, were chosen as the alternative.

Waterston's intentions in making this intervention were stated in his own description of his planned development, as noted by Niemann (2012: 101):

Suggested Development Scheme
Buy the island or persuade the National Trust for Scotland to buy it, as a Nature Reserve with the establishment of a Bird Observatory under the auspices of the Scottish Ornithological Club, with myself as resident warden & factor.

Of equal or more importance to the island in general was Waterston's intention to improve the economic situation for the residents, particularly in the context of the knitwear production and through encouraging and enabling visitors. He noted he would "come to an arrangement with all the crofters that I shall purchase their entire output of hosiery at prices to be agreed and that they must agree to do all

WATER FEATURES
BM Boini Mire
DA Da Water
ELW Easter Lother Water
FD Field Ditch
G Gilsetter
GB Gilly Burn
GW Golden Water
H Homisdale
HB Hegri Burn
KM Kirki Mire
MB Meadow Burn
MV Mire o' Vatnagard
OS Obs Scrape
SM Suka Mire
US Utra Scrape
V Vaadal
WB Wirvie Burn
W Walli Burn

MAIN OBS TRAPS
DD Double Dyke
G Gully
HD Hjon Dyke
SD Single Dyke
RS Roadside
P Plantation
V Vaadal
NG North Grind

TOILETS
FIBO
Airstrip
Hall
Stackhoull

SHOPS
Stackhoull
(*groceries
and gifts*)
FIBO (*gifts*)

KNITWEAR
Burkle
Schoolton
Upper Leogh
Nether Taft
FIBO

HOUSES
Ae Easter Houll
Ba Barkland
Br Brecks
Bs Busta
Ch Chalet
F Field
H Houll
Ha Auld Haa
K Kenaby
Ko Koolin
LL Lower Leogh
LS Lower Stoneybrek
M Midway
NS North Shirva
NT Nether Taft
Q Quoy
S Setter
Sc Schoolton
Sh Shirva
Sk Skerryholm
Sp Springfield
St Stackhoull (and Vaila's trees)
T Taft
U Utra
UL Upper Leogh
US Upper Stoneybrek

POINTS OF INTEREST
1 Heinkel crash site
2 School and Hall
3 Shop and Post Office
4 Kirk
5 Museum
6 Chapel
7 Cemetery
8 Puffinn

road/track
ditch/stream
higher areas
cliffs & intertidal
wet area/mire
water body
aerogenerator

0   250   500   1000 m

Fair Isle

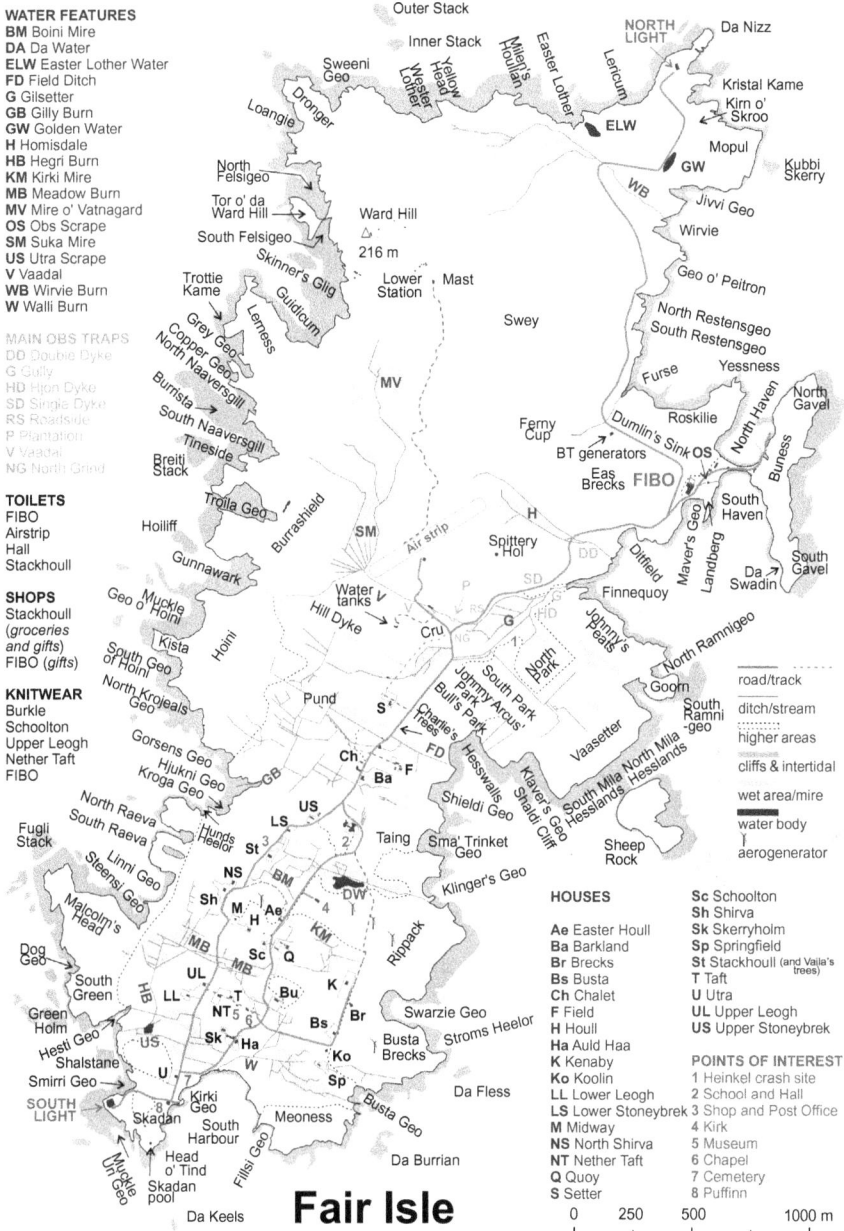

Figure 4.1 Map of Fair Isle

their marketing through me" (Niemann 2012: 102). In that way, he anticipated being able to break the monopoly of a few absentee agents who paid little for the knitwear and made large profits in reselling to major stores (Waterston 1967). That part of his plan was never fully realised but in January 1948 he did purchase Fair

Figure 4.2 The first Observatory at the North Haven (old naval huts)

Isle for £3,000 by means of a loan and began work on the renovation of the naval huts. At the same time, he undertook the construction of traps to catch birds for ringing, an essential part of the scientific work of the Observatory (over a quarter of a million birds have been ringed on Fair Isle). From the beginning, Waterston had envisaged the Observatory and his ownership of the island as being crucial for maintaining the economic viability of the community (Waterston 1967), because as noted above, the threat of evacuation was very real at that time. Waterston (1946: 114) had noted "On Fair Isle the danger stage may soon be reached when there will not be sufficient young men on the island to carry out the necessary public services, such as manning the mail boat, etc". The fact that this fate was avoided is not due to his efforts alone, but it is fair to argue that without his intervention and commitment, which led directly to many other developments noted below, the viability of the community would have remained questionable, if not impossible, over the next quarter century. The discovery and subsequent development of North Sea oil and gas reserves and their impact on Shetland (Butler and Fennell 1994; Nelson and Butler 1992) might have prevented the evacuation of Fair Isle from the late 1970s onwards, but whether the population would have stayed on the island from the 1940s until then was unlikely (Waterston 1967).

The establishment of the Observatory and the acquisition of the island combined represent the intervention being examined. It is important to note that this combined intervention worked because from the beginning the two actions had shared goals; the continued occupation of the island and the observation and recording of bird populations and migrations. The crofters on Fair Isle, particularly George Stout, with whom Waterston stayed on several visits (Figure 4.3), were skilled at bird identification and contributed greatly to the recording of species and migration.

Some of the methods used would not be accepted today, as if there was doubt about the identification of a possible new species, the bird would be shot and kept for identification. Now, fixed nets and traps along with relocatable mist nets are used for bird capture for study and ringing. The cooperation of the residents is still important in reporting the appearance of new birds and providing food and shelter for birds through agricultural practices, as well as allowing the visiting birders and Observatory staff to conduct observations in and around the crofts, including in gardens as well as the few fields of crops.

Figure 4.4 is a timeline of interventions and important events on Fair Isle from the initial birding visits before the main intervention, up to the planned next intervention (2024), the opening of a new Observatory following the total destruction of the fourth Observatory by fire in 2019 (www.fairislebirdobs.co.uk). The redevelopments of the Observatory have seen an increase in the amount of accommodation provided for visitors, as well as significant improvements in the quality of accommodation – the major change being a move from bunkhouses in the old naval huts (Figure 4.2) with ablutions, meals and social space in other separate buildings, to en suite accommodation with comfortable dining and social facilities including a library and a bar (Figures 4.5, 4.6 and 4.7). As noted below, the

Figure 4.3 George Waterston, George Stout (with shotgun) and Archie Bryson

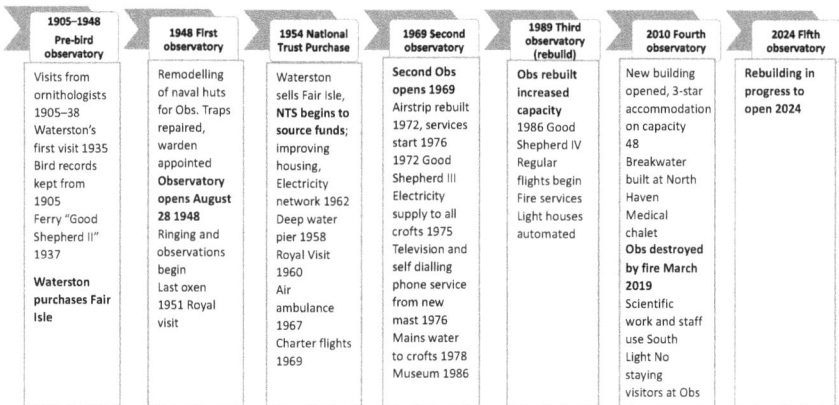

1905–1948 Pre-bird observatory	1948 First observatory	1954 National Trust Purchase	1969 Second observatory	1989 Third observatory (rebuild)	2010 Fourth observatory	2024 Fifth observatory
Visits from ornithologists 1905–38 Waterston's first visit 1935 Bird records kept from 1905 Ferry "Good Shepherd II" 1937 **Waterston purchases Fair Isle**	Remodelling of naval huts for Obs. Traps repaired, warden appointed **Observatory opens August 28 1948** Ringing and observations begin Last oxen 1951 Royal visit	Waterston sells Fair Isle, **NTS begins to source funds;** improving housing, Electricity network 1962 Deep water pier 1958 Royal Visit 1960 Air ambulance 1967 Charter flights 1969	**Second Obs opens 1969** Airstrip rebuilt 1972, services start 1976 1972 Good Shepherd III Electricity supply to all crofts 1975 Television and self dialling phone service from new mast 1976 Mains water to crofts 1978 Museum 1986	**Obs rebuilt increased capacity** 1986 Good Shepherd IV Regular flights begin Fire services Light houses automated	New building opened, 3-star accommodation on capacity 48 Breakwater built at North Haven Medical chalet **Obs destroyed by fire March 2019** Scientific work and staff use South Light No staying visitors at Obs	**Rebuilding in progress to open 2024**

Figure 4.4 Timeline of interventions

facilities of the later Observatories were open to island residents and used for weddings, anniversaries and other gatherings.

One of the reasons for the success of Waterston's interventions was his willingness to pass on the ownership of the island to the National Trust for Scotland in 1954, selling it to that charity for the same sum he paid for it in 1948. Waterston had always contemplated the involvement of the National Trust for Scotland as his comment recorded in Niemann's (2012) book showed, and although the

Figure 4.5 Second Observatory, 1969

Figure 4.6 Third Observatory, 1997

Figure 4.7 Fourth Observatory, 2014

Observatory has remained independent of the Trust, the two organisations work effectively together in the broader interests of the island and its permanent and visiting populations. Waterston deemed it necessary to sell the island because the scale of actions needed to be taken to remove the threat of evacuation proved too large and numerous for a single individual. These included building a breakwater at the North Haven harbour, cooperative production and marketing of knitwear with a registered trademark, reintroduction of weaving, improved amenities and utilities, the establishment of a museum of island life, and the use of wind power for electricity generation, almost all of which have been achieved (Thom 1989). As well, a significant number of potentially supportive agencies did not provide funds to individual property owners, meaning the change in ownership of the island to the National Trust for Scotland opened up many more avenues of financial and other assistance (Waterston 1967), as well as confirming Waterston's primary concern for ensuring population viability through a range of developments, including tourism.

Fair Isle Bird Observatory: Linkages and Developments

Figure 4.8 shows the linkages that have developed between the primary intervention, the first bird Observatory and elements of community life on Fair Isle (Butler 2019). Primary or major linkages have developed between the Observatory and tourism, employment and conservation in the context of wildlife and agriculture, as also found on other islands (Movono et al. 2017; Nyauipane and Poudel 2011).

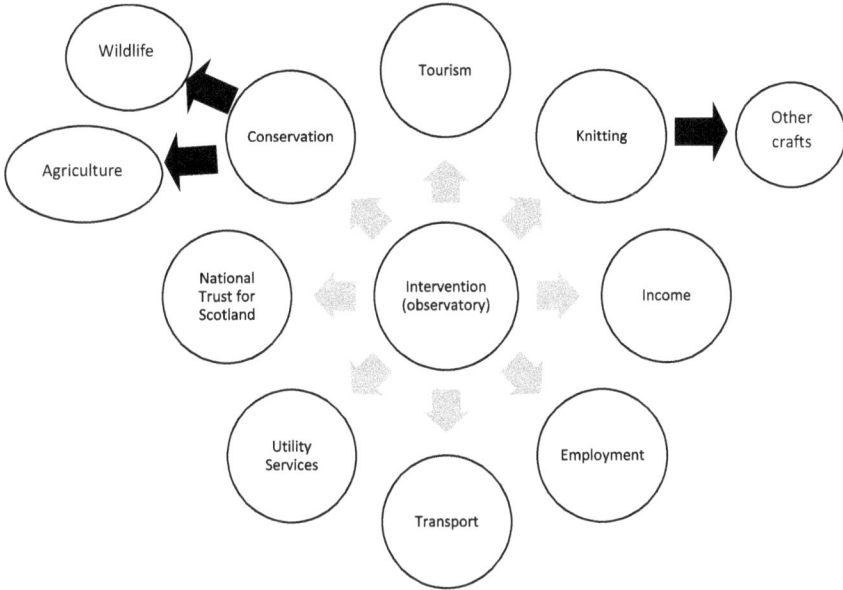

Figure 4.8 Linkages between the Observatories and the island community

Before the first Observatory was opened, there was no formal accommodation for visitors to Fair Isle. The very few ornithologists who visited had either stayed on their boats, used an unoccupied croft (Pund), or stayed with one or more crofters, George Stout at Field in particular. The opening of the Observatory provided accommodation, including meals, for an initial 12 visitors for the season that the Observatory was open, generally from April until the end of September. Each rebuilding or remodelling of the observatory increased the number of beds available for visitors (48 in the fourth Observatory), although the period of opening has remained roughly the same over the seven decades of operation. In the last two decades, some additional serviced accommodation on the island has been provided in a few (normally two or three) crofts and one self-catering former croft, adding capacity for about another ten guests at the most (Butler 2016). The Observatory in recent years has served also as a facility for individual yachts and other private boats visiting the island, with sailors being able to use the amenities of the Observatory for showers, washing and drying clothes, and also purchasing meals. In turn, the visitors have created links with other elements of island life, in particular knitting (and to a smaller extent, other handicrafts), through the purchase of products in particular, and also taking courses on the island run by knitters (Butler 2015). Tourism has also been of importance by providing another limited market for the village shop, and more importantly, by boosting passenger numbers on, and thus improving the viability of, air services and the ferry service. Visitors staying at the Observatory also visit the Museum, again providing a limited amount of income to that facility.

In terms of employment, the Observatory has normally employed a permanent warden, an administrator and other temporary staff during the season. The rangers and cook are generally being off-island employees, but islanders have a small number of positions in cleaning, maintenance and other services. As on many small islands, multiple employment is common, with many of the islanders holding several part-time positions, and thus the part-time employment at the Observatory fits in well with this practice (Figueroa and Rotarou 2016). Income is also supported by the fact that the last two forms of the Observatory had maintained a small gift shop which sold a limited number of island handicraft products, paintings, photographs, cards and knitwear. Another major economic linkage is that the Observatory purchases its food supplies through the village shop, a key element in the survival of the shop (Waterston 1967) which is of vital importance to the island population as a whole. Virtually all shop supplies are carried to the island by the *Good Shepherd IV*, the island-owned ferry boat, again slightly increasing the economic viability of the ferry operation.

There are other important social benefits to the island community from the establishment of the Observatory (Butler 2018). It serves as the only place on the island for residents, as well as staying guests, to be able to dine, and the fourth Observatory provided the first and only licensed establishment on the island. Also, the Observatory in later years became a centre for regular performances by island musicians and handicraft producers, thus strengthening relations between visitors and local residents through their interaction on those occasions. In addition, the presence of the Observatories created wider interest in Fair Isle, both amongst naturalists in general, through the observations and recordings of birds and other wildlife, and in terms of general tourists now able to visit the island for short, sometimes one-day, visits because of the air services. Tourists have also been attracted to Shetland by the novels of Ann Cleeves (a former cook at the Observatory) and the accompanying *Shetland* television series (the lead character of which is a Fair Islander), with one novel, *Blue Lightning* (Cleeves 2010) and the resulting television programme being based and filmed on Fair Isle at and around the Observatory. The Observatories have provided extra support in terms of their presence and support for other island developments, in particular the development of wind energy supply and the establishment of the air service to and from the island. The Fair Isle Bird Observatory Trust has had a positive relationship with the National Trust for Scotland since the purchase of the island by the Trust in 1954 and there has been mutual support for other projects such as the push for the waters around Fair Isle to become a marine reserve (www.fimeti.org.uk).

There are many other minor but important ways in which the Observatories, in all their forms, have proved beneficial to the island and its community, similar to other examples of the growth of tourism and its significance to small islands (Creaney and Niewiadomski 2016; Croes 2016; Ridderstaat et al. 2016). The Observatories have served a role in educating visitors, particularly those staying there, on appropriate behaviour when on the island, in regards to such aspects as respecting the privacy of residents, how and where to access fields and crops, and how to behave around sheep and other wildlife; this has been important in maintaining good relations between tourists and residents (Butler 2021), always a key issue on small islands involved in tourism (Lepp 2008). The location of the Observatories, on the north of the island (Figure 4.1), meant that they were out of sight of crofts and that staying guests were not around the permanent residences from the late afternoon onwards, providing a break for residents from the presence of tourists (Nunkoo and Ramkissoon 2010). (This location, as noted earlier, was not what Waterston had at first envisaged, but the destruction of Pund had removed his original choice, and resulted in the staying visitors being located some distance from the residents' houses.)

Visitors often assist residents with specific tasks, the most important being the gathering of sheep from the rough grazing in the north of the island and driving them to a central location for dipping, shearing and other actions. On a small island with a small population the provision of such assistance is perhaps more important than any actual benefits derived (Waterston 1967), and for many residents, the boost in population during the summer is welcomed (Kim et al. 2013). The positive attitude towards tourists is also demonstrated by the willingness to host visits from a regular number of cruise boats during the summer, when passengers can peruse a display of handicrafts for sale at the community hall and enjoy tea and home baking paid for by the cruise lines. At the present time, until the new Observatory (Figure 4.9) is opened in 2024, such visitors are compensating for the absence of staying visitors at the Observatory.

Figure 4.9 Fifth Observatory (architect's impression), 2023

Conclusions

The success of the intervention is shown by the continued permanent occupation of Fair Isle, with a population that has increased to almost double its total from the late 1940s. The National Trust for Scotland has a policy of advertising for new residents when a house becomes available (almost all properties have been modernised or rebuilt since the 1950s) and receives several hundred applicants for each vacancy from all over the world, with new island residents including English, French, American and Venezuelan citizens. Several former wardens of the Observatory have remained on the island as crofters after leaving the employment of the Observatory, reflecting the successful integration of the Observatory and its staff into the island community. Almost all of Waterston's plans and goals for the Observatory have been achieved, Fair Isle has become a significant element in the overall tourism offerings of Shetland, and the Observatory an attraction in its own right (Fennell 1996). The increasing popularity of the island with smaller cruise ships reflects a major interest in the Northern Isles from that tourist segment, with Kirkwall (in Orkney) being the most visited cruise ship port in Britain in terms of the number of ships.

It is the integration of the Observatories in their several forms with the island community, its economy and its natural environment, as well as the long-term vision that Waterston had for the island's future that explain its success. Thom (1989: 109) summed up the focus of Waterston's intent: "The successful integration of the islanders' lifestyle with conservation of landscape and wildlife … is a concept very much in line with George's dream for the island". His intervention and subsequent related developments have been social as well as economic successes, and on a small island these two elements are inextricably tied together. The Observatories have stimulated interest in, and conservation of, the natural landscape and resources of the island. It would be hard to find better examples of both sustainable development and ecotourism in practice, whereby tourism, through the Observatory, has become the 'glue' holding the island project together (Butler 2021). The fact that there is still a viable community on Fair Isle is a living tribute to George Waterston, the individual who made the initial intervention possible.

Acknowledgements

I was fortunate to meet George Waterston and have a very long conversation with him in the course of being interviewed (unsuccessfully) for the position of his assistant as representative of the Royal Society for the Protection of Birds in Scotland in 1967. That discussion covered far more than the job interview and revolved mostly around Fair Isle, the Observatory, and what it meant to Waterston. I have also been fortunate in being able to visit Fair Isle for extended periods in 1962 and 1963, and again in 2012 and 2013 to redo the research undertaken in the first of those periods thanks to support from the Leverhulme Trust. I am grateful to the many islanders for their hospitality on my eight visits in total, to Peter Davis (a fellow geographer) the warden in 1962, and to officials of the National Trust for

Scotland and Shetland Island Council for their assistance over the years. Finally, I am particularly grateful to the Fair Isle Bird Observatory Trust for allowing me to include photographs from their archives to illustrate this chapter.

References

Alberts, A. and Baldachinno, G. (2017). Resilience and tourism in islands: Insights from the Caribbean. In Butler, R.W. (Ed.), *Tourism and Resilience*. CABI, Wallingford, pp. 150–162.

Amoamo, M. (2017). Resilience and tourism in remote locations. In Butler, R.W. (Ed.), *Tourism and Resilience*. CABI, Wallingford, pp. 163–180.

Briguglio, L. (1995). Small island developing states and their economic vulnerabilities. *World Development*, 23(9): 1615–1632.

Butler, R.W. (1963). *A Geographical Study of the Development of an Isolated Island Community.* Unpublished B.A. thesis, University of Nottingham, Nottingham.

Butler, R.W. (2014). Understanding the place of birdwatching in the countryside: Lessons from Fair Isle. In Dashpter, K. (Ed.), *Rural Tourism: An International Perspective.* Cambridge Scholars Publishing, Cambridge, pp. 321–336.

Butler, R.W. (2015) Knitting and more from Fair Isle, Scotland: Small-island tradition and micro-entrepreneurship. In Baldachinno, G. (Ed.), *Entrepreneurship in Small Island States and Territories*. Routledge, Abingdon, pp. 83–96.

Butler, R.W. (2016). Changing politics, economics and relations on a small island, *Journal of Island Studies*, 11(2): 687–700.

Butler, R.W. (2018). Niche tourism (birdwatching) and its impacts on the well-being of a remote island and its residents. *International Journal of Tourism Anthropology*, 7(1): 5–20.

Butler, R.W. (2019). Fair Isle: Half a century of change. *Scottish Geographical Journal*, 135: 125–138.

Butler, R.W. (2021). The potential impacts of climate change on avitourism to islands: The case of Fair Isle, Scotland. *International Journal of Islands Research*, 2(1): 27–40.

Butler, R.W. and Fennell, D.A. (1994). The effects of North Sea oil development on the development of tourism, *Tourism Management*, 15(5): 347–357.

Cleeves, A. (2010) *Blue Lightning.* Pan Macmillan, London

Creaney, R. and Niewiadomski, P. (2016). Tourism and sustainable development on the Isle of Eigg, Scotland. *Scottish Geographical Journal*, 132(34): 210–233.

Croes, R. (2016). Connecting tourism development with small island destinations and with the well-being of the island residents. *Journal of Destination Marketing and Management*, 5(1): 1–4

Duncan, A. (ed.) (1938). *A Bird-Watcher's Diary by Mary, Duchess of Bedford.* Printed for private circulation. John Murray, London.

Fair Isle Marine Conservation Area (n.d.) http://www.fimeti.org.uk (Accessed 6 August 2017).

Fennell, D.A. (1996). A tourist space-time budget in the Shetland Islands. *Annals of Tourism Research*, 23(4): 811–829.

FIBO (Fair Isle Bird Observatory) (n.d.). *Annual Reports*. Fair Isle Bird Observatory Trust, Swallowtail Print, Norwich. Published on website, https://www.fairislebirdobs.co.uk (Accessed 16 May 2017).

Figueroa, B. and Rotarou, E.S. (2016). Tourism as the development driver of Easter Island and the key role of resident perceptions. *Island Studies Journal*, 1(1): 245–264.

Kim, K., Uysal, M. and Sigg, M.J. (2013). How does tourism in a community impact the quality of life of community residents? *Tourism Management*, 36(4): 527–540.

Lepp, A. (2008). Attitudes towards initial tourism development in a community with no prior tourism experience: The case of Bigodi, Uganda. *Journal of Sustainable Tourism*, 16(1): 5–22.

Midgley, G. (2000). *Systemic Intervention Philosophy, Methodology, and Practice.* Springer, Heidelburg.

Moretti, S. (2021). State of the art of cultural tourism interventions. Deliverable D3.1 of the Horizon 2020 project SmartCulTour (GA number 870708), published on the project website in May 2020, http://www.smartcultour.eu/deliverables (Accessed 21 December 2022).

Movono, A., Dahels, H. and Becken, S. (2017). Fijian culture and the environment: A focus on the ecological and social interconnectedness of tourism development. *Journal of Sustainable Tourism*, 26(3): 451–469.

Nelson, J.G. and Butler, R.W. (1992). Assessing, planning, and management of North Sea oil effects in the Shetland Islands. *Environmental Impact Assessment Review*, 13(2): 201–227.

Niemann, D. (2012). *Birds in a Cage.* Short Books, Exeter

Novelli, M. and Benson, A. (2005). Niche tourism: A way forward to sustainability? In Novelli, M. (Ed.), *Niche Tourism: Contemporary Issues, Trends and Cases*. Butterworth Heinemann, Oxford, pp. 247–250.

Nunkoo, R. and Ramkissoon, H. (2010). Community perceptions of tourism in small island states: A conceptual framework. *Journal of Policy Research in Tourism, Leisure and Events*, 21(1): 51–65.

Nyaupane, G.P. and Poudel, S. (2011). Linkages among diversity, livelihood and tourism. *Annals of Tourism Research*, 38(4): 1344–1366

Ridderstaat, J., Croes, R. and Nijkamp, P. (2016). A two-way causal chain between tourism development and quality of life in a small island destination: An empirical analysis. *Journal of Sustainable Tourism*, 24(10): 1461–1479.

Scott, D. and Thigpen, J. (2003). Understanding the birder as tourist: Segmenting visitors to the Texas hummer-bird celebration. *Human Dimensions of Wildlife*, 8(3): 199–218.

Shetland Island Council (SIC) (2012). Discussion with SIC staff (anonymity preserved). August 2012, Lerwick.

Thom, V. (1989). *Fair Isle: An Island Saga.* John Donald Publishers, Edinburgh.

Waterston, G. (1946). Fair Isle. *Scottish Geographical Magazine*, 62(3): 111–116.

Waterston, G. (1967). Personal communication.

Willis, D.P. (1967). Population and economy of Fair Isle. *Scottish Geographical Magazine*, 83(2): 113–117.

5 Commodifying Landscape to Conserve It?

The Politics of Highest and Best Use of Land in Indonesian UNESCO Geoparks

Rucitarahma Ristiawan, Edward Huijbens, and Karin Peters

Introduction

In 2019, the Indonesian government published a Presidential Regulation (No 9/2019) regarding the development of geoparks in Indonesia. This policy responded to the growing urgency to regulate and centralize the administration of the already existing geoparks' development initiated at regional and local government levels. The success of geopark development at the regional level and the shifts in power this entailed prompted the government to develop a regulatory framework in order to wrest back control of the country's development. The regulation also details instructions for developing geoparks, identifying these as a national economic alternative resource and as broadening the revenue base of the national economy. Several Indonesian ministries responded to this regulation by proactively creating programs in infrastructure development, and land use conversion to conservation, zonation of conservation areas and creation of investment platforms to accrue benefits from the conservation of the geological landscape as per the geopark aims (Ristiawan et al. 2023a). The consolidation of geopark development at a national level responded to governments at local and regional levels, having established networks and business alliances involving local elites and private investors for building road infrastructure and accommodations to support tourism growth and to secure their power in land tenurial control. The identification of geo-heritage conservation as the highest and best use of land, and its subsequent framing by different levels of government, consequently brought noticeable impacts on the practices of land use and management by local communities, leading to changes in their socio-economic practices. In this chapter, we illustrate these particular dynamics by tracing land selling and acquisition for tourism purposes and the ways in which these effectively commodify the landscape under the guise of conservation.

The rent gap is the conceptual framework adopted for understanding the process of valuing and commodifying the landscape. Smith (1979) explains that the rent gap is the gap between land and property's actual and potential value in an urban context. Two key concepts were coined by Smith (1979, 1987) to explain how rent gaps work – 'potential ground rent' and 'actual ground rent'. Potential ground rent

DOI: 10.4324/9781003449027-6

refers to the income landlords receive under the land's highest and best-use scenarios, whereas actual ground rent refers to the income landlords currently receive for land use, if any (Smith 1987; Slater 2017). Two well-documented strategies to widen the gap and thereby increase the potential rent possibly accrued by the landholder are devalorizing through disinvestment in a parcel of land or property (Smith 1979) and placing a new projection of the highest and best value of land through legal regulation (Slater 2017).

In this chapter, we analyze the relationship between geopark tourism development strategies and landscape commodification, unravelling who benefits from the establishment of the geoparks, how different interventions of landscape development rely on the creation of a rent gap, and how these interventions contribute to the development of the destination and the community. We will do so by looking into two cases in Indonesia: Gunung Sewu Geopark and Ciletuh-Palabuhanratu Geopark. These cases represent different practicalities and processes of implementation of the geopark program. We explore the processes by which this highest and best use is created by bringing the rent gap theory into conversation with the political economy of tourism. The discussions taking place in this chapter are part of a broader doctoral study and are based on the empirical insights generated through grounded fieldwork in communities around the two parks in question. We show that the making of tourist places framed as geoparks and discussions around geo-heritage conservation at regional and local levels relies on expanding the rent gap, redefining and widening local inequalities, and enhancing landscape vulnerability meant to be overcome by the geopark development program. This process of landscape commodification subsequently engendered responses from the community to accrue benefits from their land, which correlates with voluntarily surrendering their means of production. These insights are crucial for implementing the presidential regulation from 2019 and should be considered when consolidating geopark strategies at a national level.

Rent Gap in the Political Economy of Tourism

In 2010, David Harvey argued in *The Enigma of Capital and the Crises of Capitalism* that rent has to be brought to the forefront of analysis to understand the ongoing production of space resulting from the circulation and accumulation of capital. Addressing this in a tourism context, Yrigoy (2023) frames the political economy of tourism (PET) around the physical space where tourism takes place and sees the grounded process of place-making as a critical component of the broader tourism commodity. Yrigoy's forefronting of place corresponds with ongoing efforts to extend the conceptualization of tourism space beyond being simply an expression of the commodification of the social time for recreation (see MacCannell 1976; Mosedale 2006). Yrigoy (2023) argues that tourism commodification goes beyond tourism experiences and encompasses the built environment in making the spaces and places of hospitality and tourism. By including these contextual features of destination development, a better understanding of how the expanded circulation of capital within tourism operates is provided.

Echoing this, the underlying conceptual framing on this chapter is the rent gap, which has emerged as a prominent tool to understand the capitalist production of space and has occupied the attention of several scholars in tourism geographies (see Kulusjärvi 2020; Young & Markham 2020; Huijbens & Jóhannesson 2019). The most prominent adoption of the rent gap ideas is in urban settings, and most particularly around Airbnb development (see: Yrigoy 2019; Amore et al. 2022). Slater (2017) criticized rent gap scholars for using rent gap theory predominantly to understand capitalist space production in contemporary urban studies. He furthermore argues that rent gap theory helps to expose and confront new geographies of structural violence more generally, including analyzing the power of land developer interests in processes of capital accumulation globally. Given the particularity of tourism destination development in a non-urban context, we then propose to establish a dialogue between Smith's (1979) rent gap theory and the commodification of landscapes through tourism, framed within the context of the political economy of tourism (Yrigoy 2023). Hereby, we bring the capitalist production of space that creates avenues to accommodate capital accumulation into dialogue with tourism through Smith's (1979) formulation of the rent gap.

In Indonesia's geopark development context, this dialogue provides a framework to understand how landscape conversion and sustainable tourism development can play out differently for actors involved. On the one side, the rent gap fruitfully enriches tourism destination studies, bringing rent as the particular lens of seeing how tourism transforms places into destinations. On the other side, the political economy of tourism adds particular insights into the community's adaptive strategies for coping with processes of dispossession resulting from actors capitalizing on the rent gap created by the tourism development process. As community empowerment through tourism destination development is partly the focus of sustainable tourism studies (Carr et al. 2016), these considerations are extended through the rent gap by exposing the structural violence within the production of tourist places (Darling 2005). The constitutive power of tourism development can then be understood as part of a larger, even globalized, capitalist development (Britton 1991).

The Political Economy of Tourism Rent in the Making of Tourism Destinations

By focusing on the process of capital accumulation through rent and the formation of rent gaps as places are transformed into tourism destinations, we want to show how the commodification of space works within tourism. Tourism spaces are converted into commodities in several ways (Tomassini & Lamond 2022). One of these ways is the physical transformation of a place into a tourism commodity (Young & Markham 2020). Once the land itself has gained value and transformed, tourists are able to use these spaces, and tourism revenues can be created from the tourists purchasing access to the land and the commodities, services and experiences built and offered on the land (Bianchi 2018).

As elaborated above, Yrigoy's (2023) point of departure was that the physical transformation of space is crucial for tourism commodification to take place.

By taking this approach, land rent thereby appears as the cost of production for new capital investment in a tourism context. However, Yrigoy's (2023) conception of land rent as the cost of production and, at the same time, as potential revenue omits the detailed mechanisms of governing land and regulating market transactions. What sort of mechanism and deals need to take place to allow the land itself to be accounted as costs, and at the same time, what are the benefits identified by the multiple actors? In other words, under what conditions can the landscape be commodified, and what sort of political dynamics inform the decisions on the highest and best use of the landscape? This process of negotiating the highest and best use of land is the process that transforms the position of land as the means of production (constant capital) into a fictitious commodity (Polanyi [1944] 2001). This transformation signifies the opening of the rent gap (Smith 1979; Slater 2017; Krijnen 2018). Eventually thus, the highest and best use is a policy product that accommodates capital accumulation for some and not others (Harvey 2010).

Benefitting from Yrigoy's (2023) conceptions but adding these considerations on the actual mechanisms of valuing, we move from the focus on how the landscape has been transformed into questioning the underlying mechanism of why specific landscapes have been selected as sites for transformation and the negotiations around them. In this sense, the meaning of the 'highest' and 'best' land use becomes revealed as contested and negotiated (Daugstad 2008). In examining this, we centre our analysis on the dynamic of landowners who occupy specific importance in the landscape through privatization and capitalization. To this extent, the active practices of surrendering the land by certain landowners as their means of production cannot be detached from the larger socio-cultural-economic structure hegemonically campaigned for within specific contexts. In Indonesia, the surpluses made from land selling do not necessarily transform rural landowners into capitalists (Batubara 2023). Many of those selling land do so to gain access to, and include themselves in, state institutions by investing their surpluses for their families' education in the cities, thus setting them up for working in monthly waged public sector jobs (Lund 2023). This elaboration echoes what Harvey (2010) states – that the rent itself is a political avenue and that the decision to surrender the potential to accrue rent is made in particular contexts.

In this sense, we use the rent gap not to tease out the mechanism of how the actual ground rent is being created but instead to see how the projection of potential rent becomes part of a hegemonic discourse to accumulate capital in specific ways. In doing so, we perceive rent within the highest and best-use framework as something that has been designed through particular land governing practices. We hereby focus on how land governance in a particular context shapes and can be shaped by diverse actors' willingness to accrue the potential value of the land. This willingness to accrue land's potential value will create diverse forms of capital accumulation, not only by the government and local elite actors but also by the community previously assumed to be less powerful and marginalized. In this sense, attempts to accumulate capital appear not only from the business and investment sector but also from particular types of landowners as they surrender their means of production.

The Production of the Highest and Best Use in the Gunung Sewu Geopark

Gunung Sewu Geopark is located approximately 30 km east of Yogyakarta, the capital city of Yogyakarta Special Region Province on Java Island. It stretches 85 km west-east from Parangtritis Beach of Yogyakarta Special Province to Teleng Ria beach of Pacitan, East Java. It covers 180,200 ha and is home to approximately 805,000 people. It is under the auspice of three provincial governments: Yogyakarta Special Region, Central Java and East Java. These three provincial governments manage the total of 33 registered geosites in this geopark. It obtained its UNESCO Geopark Status in 2015.

The decentralization policy introduced in Indonesia after the country's 1998 political turmoil enabled regional governments to have optimal power in regulating land use development and investment in Indonesia (Bakker & Moniaga 2010; Kurniadi 2020). In sustaining this, regional governments needed to seek independent sources of income with minimum subsidy from the central government to fund their regional development programs. In the Gunung Sewu area, the authority taking charge of the development in the region is the Yogyakarta Sultanate. More than 70% of the Gunung Sewu area falls under the Yogyakarta Special Province, led by the Sultan of Yogyakarta as the governor. Historically, most of the land in the coastal area is owned by the Yogyakarta Sultanate supported by Gubernatorial Regulation (Pergub No. 11/2008), which regulates explicitly how village land achieves the status of 'Crown Domain'. The Sultan of Yogyakarta was also granted a lifetime status as governor of Yogyakarta by the national government through the National Act (No. 13/2012), giving him the power to control all land use and transactions, supporting the royal family business (Kurniadi 2020).

In regard to Gunung Sewu, the identification of the highest and best use of the land looked to what was happening in Bali in early 2000 as it emerged as the centre of Indonesian tourism. The mission of creating a 'new Bali' in Yogyakarta was initiated and to sustain this mission the regional government claimed that the existing agriculture-related land use did not suit the regional objectives to generate independent income. Even though the community used the land in the coastal area for small-scale cassava and *jati* (*Tectona grandis*) plantations, the regional government and the Yogyakarta Sultanate did not perceive those existing land uses as giving the best return of investment for the region.

Moreover, knowledge generated on the uniqueness of the karst landscapes in the Gunung Sewu area from studies by Universitas Gadjah Mada during the early 2000s through to 2010 (see Haryono 2000; Haryono 2001; Haryono & Suratman 2010) became a particular consideration for the regional governments to gain public trust to further their development plan. They saw an opportunity to conserve the geological landscape and, at the same time, make use of the landscape as the source of regional income through tourism. Later, these studies became the grounding point of the Energy and Mineral Resource Ministry Decision Letter (No 3045 K/40/2014) about Gunung Sewu Karst Area Protection. At that stage, the geopark concept was selected to frame this protection. Practically, applying the geopark

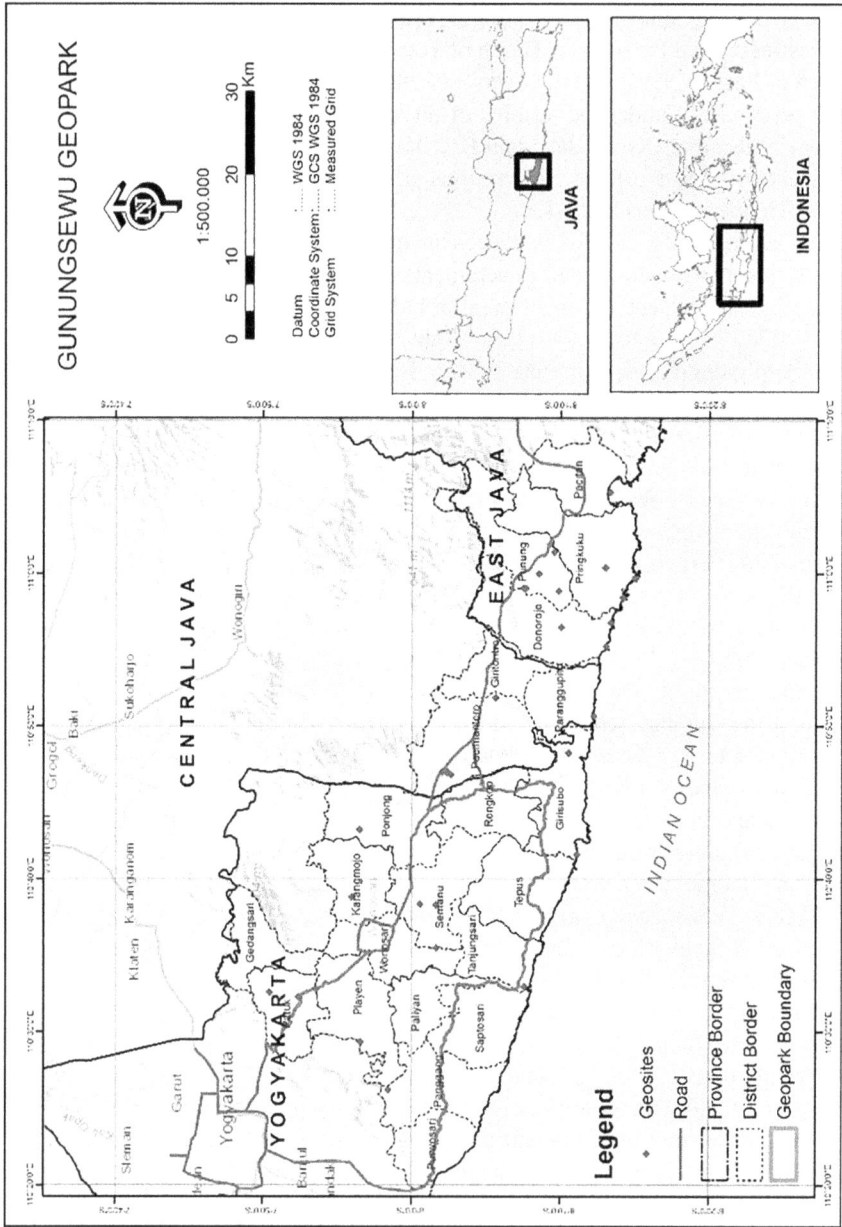

Figure 5.1 Location of Gunung Sewu Geopark

Source: Ristiawan et al. (2023b) (reused with permission)

concept was seen as a vehicle to accommodate the need to develop tourism through investment projects in accommodation and infrastructure, without violating the national law on karst landscape conservation as stipulated in the decision letter.

However, the landowners of most of the coastal area in Gunung Sewu, the Yogyakarta Sultanate and the regional government only had the authority, not the money, to establish infrastructure and services. To transform this land into a resource to accumulate capital, the Yogyakarta Sultanate provided an exclusive investment platform offering guarantees of stability and security for any investors who wanted to establish a tourism business on their land. The political and economic relations of this transformation are not challenging to disentangle. Owners of the villas and resorts in the Gunung Sewu coastal area are families and business partners of the Yogyakarta Sultan, owning more than 90% of the tourist resorts and accommodation businesses. Indeed, to these wealthy urbanites, locally called '*Wong Gede*' ('The Big Man') (Ristiawan et al. 2023a), the coastal area of Gunung Sewu became a highly desirable place to expand their business, especially with the protection from the Sultan of Yogyakarta.

One of the examples of a scheme initiated to protect the investment program was published in 2016. That year, the Regent of Gunungkidul issued a wholesale reform of the accommodation business permitting process by reducing the strictness of legal permits and making licencing feasibility studies subject to owners' discretion, which effectively made the environmental impact assessment optional. These schemes facilitated another spurt in tourism business development from 2016 to 2021, predominantly determined by the newly established rest areas, restaurants and private accommodations. As a consequence, agricultural land further decreased from 30,000 ha in 2016 to 22,000 ha in 2021. Most of this land was converted to housing, tourism accommodations and restaurants supported by the relatively easy business permit processes. Moreover, the Gunung Sewu regional government informally issued permits for establishing and operating resorts to speed up the establishment of accommodation businesses. In this informal procedure most resorts did not perform an AMDAL (Analisis Mengenai Dampak Lingkungan/Environmental Impact Analysis), which is compulsory from the national government to attain a legal certificate to operate a business, especially regarding green field development. Instead of sanctioning these resorts, the regional government allowed businesses to replace the AMDAL with self-sourced acclaimed scientific papers – for example, a company's independent feasibility study to justify the feasibility of the business concerning their environmental impact. The investment platform and the specific licencing platform became avenues for negotiating the practices of accruing land potential value. Tourism appeared to be the best option with Bali as a role model and geoparks emerging as the framing concept as the landscape was identified as unique. Given that investing in the agriculture and mining sector was perceived as more costly and having higher risks for investors, geoparks prevailed.

Responding to the hegemonic geopark tourism campaign, local community landowners started to perceive the potential benefits of selling their parcel of land to a tourism investment program. The communities in Girikarto, Kanigoro and Djepitoe Village mostly took advantage of this. Land selling became easier for them since the demand for their land increased. For them, selling the land to the private sector and/or regional government became an option to earn surplus money for funding their family education to prepare them to work in the state sector.

The logic of land rent becomes particular in those three villages. Instead of capitalizing on the potential rent being generated by tourism investments directly through getting paid by renting the land to an investment project, selling the land to the investment project was perceived as more beneficial. This land-selling practice also comes with a deal that the communities can work in the tourism business once the business from the government/private sector is established. This deal between communities and investors became more common, at least in the coastal areas of Girikarto, Kanigoro and Djepitoe Village, with a general assumption by the communities that double benefits from selling the land can be extracted: money from land selling and wages from working in the tourism business settled on their former land.

However, the built tourism infrastructure changes the landscape's appearance, affecting the soil structure and formation, and making it vulnerable to land subsidence. The land use change resulting from tourism investment directly caused flooding and land subsidence in other parts of the area where less intensive tourism development occurs. During November–December 2022, the first author counted more than 20 land subsidence events happening in the Gunung Sewu area. The land subsidence occurs because of the inability of the karst area on the coastal side to catch rainwater, resulting in flooding towards the central area of Gunung Sewu, which has a lower elevation. The coastal area's failure to absorb the rainwater resulted from a large-scale deforestation to accommodate resorts and tourism infrastructures and attractions. Paradoxically, it seems that geopark developments to conserve the unique landscape are resulting in its actual degradation with those not at the centre of development and able to sell their land, losing it.

Tourism Development and the Creation of Highest and Best Use in Ciletuh Geopark

Ciletuh-Palabuhanratu Geopark lies in Sukabumi region, West Java Province. Home to 25,684 residents, it covers 126,100 ha or approximately 30% of the Sukabumi region's total area, consisting of eight districts and 74 villages. It consists of 30 geosites and was granted its UNESCO Geopark Status in 2016.

The geopark initiative emerged from the brainstorming of Biofarma LLC (an Indonesian state-owned enterprise), academics from Padjadjaran University and PAPSI (a local conservation organization) about creating an alternative source of income to agriculture for the community in 2013. This was framed as a program of corporate social responsibility (CSR) for Biofarma LLC and involved the assessment of geotourism potentials in the Ciletuh area and a community-based approach to tourism development. During 2013–2015, micro homestay settlements were developed by the local communities in this spirit. The work towards obtaining a geopark status engendered attention from the tourism market, which led to an increase in number of visitors between 2013 and 2015 by more than 400%.

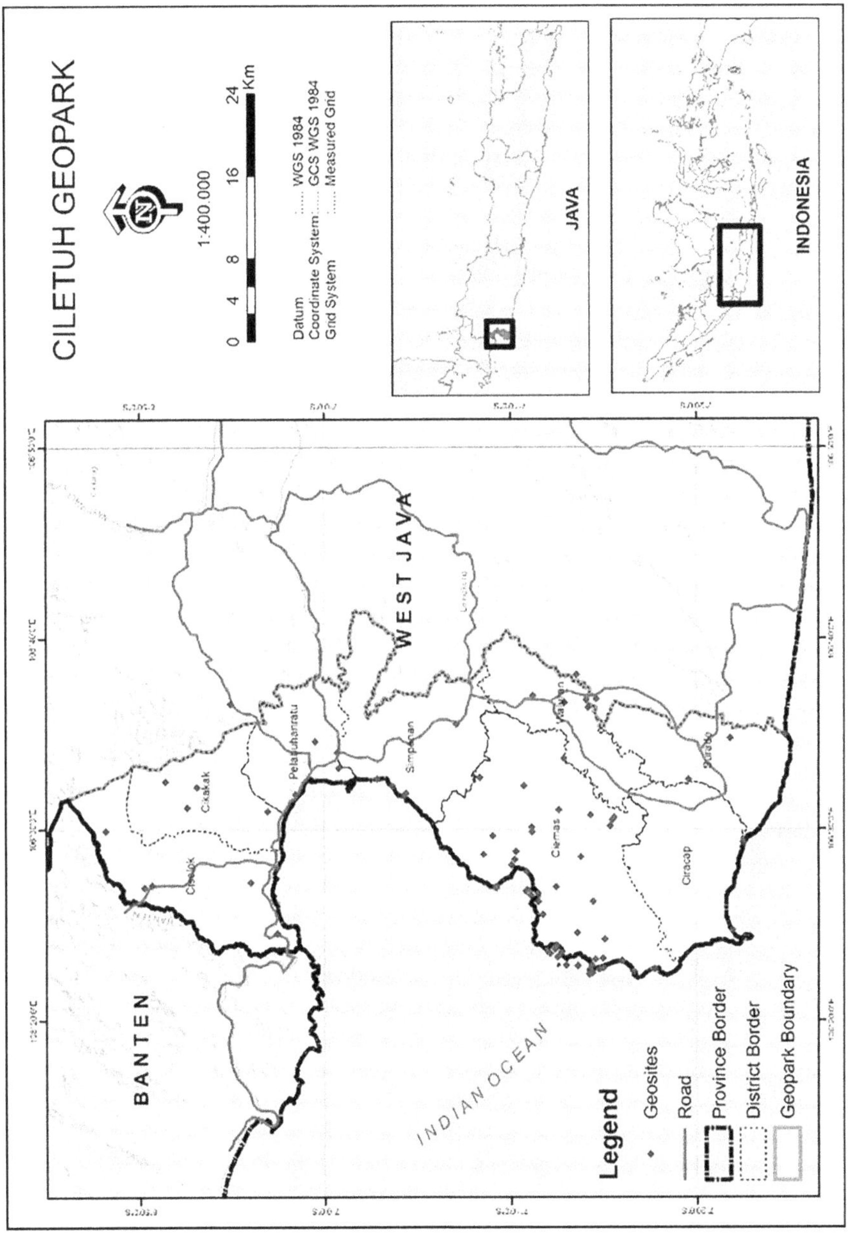

Figure 5.2 Location of Ciletuh-Palabuhanratu Geopark
Source: Ristiawan et al. (2023b) (reused with permission)

Realizing the potential economic benefits from land use change to tourism through the UNESCO Geopark Program, the village government officers saw the geopark as an opportunity to claim several areas as village land which were previously claimed by 'Balai Konservasi Sumber Daya Alam' (National Centre of Nature Reserve Conservation), then called BKSDA. They used the Indonesian agrarian land reform act of 1960 to justify their land claims. For the village government, land control became a cemented way to secure political power and gain more trust from the people.

At the same time, access to land control became a power resource to regulate and establish new businesses on the land. One of the earliest investment programs was from the ex-governor of West Java Province, who made a deal with the village head to develop a resort and paragliding site, with the provision that 5% of the profits should go to the village and the business should employ local community members. The deal directly changed the perception of the highest and best use of the land in the Ciletuh area, with several notable political actors in West Java Province – for example, the head of Sukabumi regional government, the head of Ciemas District and several regional parliament members – joining the business alliance of the ex-governor and the Ciwaru Village head. Most of them are actively involved in the same political party.

The establishment of large-scale resorts and accommodations (more than 5 ha) mainly took place in the ex-BKSDA areas claimed by the village government. Instead of renting the place to the private sector, the village government and the local elites preferred to sell the land and extract benefits differently. The common practice to extract benefits was selling the land under the condition that the newly established business should pay a monthly security fee and 'voluntary funds' to the village government. The voluntary fund is a framing of illegal fees being asked for by the village government from the private sector when they suddenly need money for their household/office needs. The village government and local elites perceived renting the land and extracting direct rent from it as less beneficial than what they call '*Politik balas budi*' (The politics of returning the favour). This politics means that the village government can ask for any favour from the private sector to return the favour of helping them settle their business.

In Ciletuh, the negotiation of the highest and best use also affects communities with small-scale land ownership. Seeing the capitalized land nearby, more communities in the coastal area, such as in Ciwaru and Girimukti Villages, started to build small-scale accommodations such as homestays and small villas using a small parcel of land previously used for small-scale rice field farming. Local community members have thus established at least 130 homestays by converting their farm fields to tourism-related use since 2017. To manage the small-scale homestay business, the community, the village government and the alliance also established a homestay organization for the locals. This organization attempts to control the number of homestays, regulate the service standard and ensure the inclusion of locals. Subsequently, homestay owners must pay a particular fee to the village government for the establishment, security fund and '*sumbangan sukarela*' (voluntary fund). Interestingly, this organization later became the political party voice supporting the village government head and his alliance in any election.

Different actors thus seek to benefit from the use of land for a particular purpose. For the landowners from the village government and local elite side, controlling and governing the land deals generates benefits and income from different types of sources: income from land selling, security funds and the voluntary funds provided by tourism business owners. The case is slightly different with landowners from the local community; they only have homestay and small-scale tourism businesses. Even though they support tourism development in the area, they only get income from tourists purchasing their accommodation services. Local communities in the coastal areas perceive that remaining in the agriculture sector does not give them the same benefits as shifting their occupation to the tourism sector. This situation shows a new form of community dependency towards the tourism sector.

This dependency on tourism creates a precarious situation for the less powerful landowners. These landowners have no power to control the direction of development, meaning that their income strongly depends on the policy determined by the village and the local government assemblage. The COVID-19 pandemic showed how vulnerable they are, as more than 70% of homestay owners decided to close their homestay businesses and leave the area to find other work in neighbouring areas. Some of them were also involved in the illegal trafficking of some cultural heritage artefacts found on several sites (Smith et al. 2022). Furthermore, the infrastructure built for tourism also engenders land instability, resulting in some land subsidence events in Ciletuh. In November–December 2022, at least seven events of land subsidence damaged the road infrastructures built for the UNESCO Geopark in 2017 in Ciwaru, Tamanjaya and Girimukti villages, affecting more than 50 farming households as they were not able to distribute their farming yields to the nearby big cities.

Discussion and Conclusion

We have discussed two specific tourism development projects in two geopark programs gaining official UNESCO geopark status in 2015–2016: Gunung Sewu, a government-led initiative geopark project; and Ciletuh, a community-led geopark program initiative. In both cases, we have shown that the highest and best use of land for tourism comes about through local/regional governance negotiations and manoeuvres to accrue capital. Diverting from the state-induced rent gap opening (Slater 2017) and market-induced rent gap opening (Krijnen 2018), the agreement on the negotiated highest and best use of land in the two cases came without clear scientific and policy justification on the most appropriate land use. Backdoor agreements characterized by clientelist practices were the dominant approach in negotiating land use and deals, mainly done through networks consisting of regional government officers, local elites and investors. Government officers' ownership of private business accommodations has become common practice in Indonesia since the New Order era starting in the late 1960s (Batubara et al. 2018). Moreover, as identified by Kurniadi (2020) and Bakker and Moniaga (2010), Indonesia's decentralized governance system offers insight into how the existing patrimonial governance tried to protect and ensure community welfare and create constant revenue from their cooperation with the private sector through establishing business enterprises. This governance system continuously creates land acquisition and dispossession through different economic modalities, the most recent one being tourism. The creation of assurance through this patrimonial structure, in this sense, works as a control mechanism of benefit sharing between stakeholders to accrue land's potential value. At the time of writing, the national government is trying to gain more control over the development processes and the presidential regulation of 2019 needs to be viewed in that light.

Even though rent was being extracted differently by landowners, landowners from the local government and local elites were the main beneficiaries of this transformation to tourism. In contrast, the landowners from the local community only got income from tourists purchasing their services. However, there are local community landowners who sell their land and do not use the surplus money from the sale to expand themselves as agrarian capitalists or tourism investors. Instead, people try to ensure themselves a future, using the money to fund their family's education in the expectation they will become state apparatus or work in the formal public sector. This is common practice in Indonesia, where people perceive surrendering themselves to the state as the most reasonable and logical practice for staying alive (Li & Semedi 2021).

As also happens in other forms of land dispossession in Indonesia, e.g., in water infrastructure (Batubara 2022), road transportation infrastructure (Sloan et al. 2018) and production forestry (Peluso & Vandergeest 2020), the highest and best use of land through tourism entails more effects for citizens with low incomes than for the ones with higher incomes. In our case, the networks creating a dispossessed community appear slightly different. Instead of becoming the extension of the state's hand in land acquisition practices, these regional networks carried out acts of resistance towards the state by assembling power in the local context to manage the land deals. This manifestation of Indonesian decentralization policies post-1998 has produced a positive trend of capital accumulation through using rent as the tool to produce safe spaces for investments in accommodation, but is also directly connected to the production of precarity and vulnerability in the same place for those with lesser means.

Through the differentiated environmental and social impacts of rent gap formation detailed in this chapter, we see that the larger the rent gap, the more precarious the condition becomes for community members of lesser means. Transforming a place into a tourism destination by capitalizing and privatizing the land will always result in environmental and socio-cultural issues affecting local populations, contradicting the principle of sustainable tourism. Since tourism projects contribute to widening the rent gap, it is time to reflect on the political economy of tourism rent, not solely using the framework to explain the situation but using the framework to foreground the emancipative approach of creating inclusive tourism destinations and dismantling the rent gap (Slater 2017). Additionally, tourism development is always relational. This means that examining tourism can focus not only on how capital is transferred to the place through rent gap creation and projection of the highest and best use of land, but also on the cause and effect of precarity in livelihoods both in situ and ex situ. Rather than imposing tourism as the sole strategy in developing communities, we should seek ways to support existing livelihoods through tourism whilst recognizing the complicated and contextual nature of tourism destination development. Most importantly, we must look back and theorize what sort of revolutionary potential tourism can have.

References

Amore, A., de Bernardi, C. & Arvanitis, P. (2022). The impacts of Airbnb in Athens, Lisbon and Milan: A rent gap theory perspective. *Current Issues in Tourism*, *25*(20), 3329–3342.

Bakker, L., & Moniaga, S. (2010). The space between: Land claims and the law in Indonesia. *Asian Journal of Social Science*, *38*(2), 187–203.

Batubara, B. (2022). Crisis, injustice, and socio-ecological justice. *IndoProgress*, *2*(1), 71–102.

Batubara, B. (2023). What revolutionary theory do we have? Reaction to Christian Lund's article at the closing session of 2023 IOS Fair Transition/LANDac International Conference, Utrecht, 30 June 2023. Retrieved from: https://forestcity.sites.uu.nl/2023/07/09/reaction-to-christian-lunds-article-at-the-closing-session-of-2023-ios-fair-transition-land ac-international-conference-utrecht-30-june-2023/.

Batubara, B., Kooy, M. & Zwarteveen, M. (2018). Uneven urbanisation: Connecting flows of water to flows of labour and capital through Jakarta's flood infrastructure. *Antipode*, *50*(5), 1186–1205.

Bianchi, R. V. (2018). The political economy of tourism development: A critical review. *Annals of Tourism Research*, *70*, 88–102.

Britton, S. (1991). Tourism, capital, and place: Towards a critical geography of tourism. *Environment and Planning D: Society and Space*, *9*(4), 451–478.

Carr, A., Ruhanen, L. & Whitford, M. (2016). Indigenous peoples and tourism: The challenges and opportunities for sustainable tourism. *Journal of Sustainable Tourism*, *24*(8–9), 1067–1079.

Clark, E., & Pissin, A. (2020), Potential rents vs. potential lives. *Environment and Planning A: Economy and Space*, *55*(6), 1–21.

Darling, E. (2005). The city in the country: Wilderness gentrification and the rent gap. *Environment and Planning A*, *37*(6), 1015–1032.

Daugstad, K. (2008). Negotiating landscape in rural tourism. *Annals of Tourism Research*, *35*(2), 402–426.

Harvey, D. (2010). *The Enigma of Capital and the Crises of Capitalism*. London: Profile.

Harvey, D. (2017). *Marx, Capital, and The Madness of Economic Reason*. Oxford: Oxford University Press.

Haryono, E. (2000). Some properties of epikarst drainage system in Gunung Kidul Regency, Yogyakarta, Indonesia. *The Indonesian Journal of Geography*, *32*(2000).

Haryono, E. (2001). Nilai hidrologis bukit karst. *Sumber*, *10*, 20–26.

Haryono, E., & Suratman (2010). Significant features of Gununngsewu Karst as geopark site. In *4th International UNESCO Conference on Geoparks*, Langkawi Geopark (Malaysia), pp. 1–9.

Huijbens, E. H., & Jóhannesson, G. T. (2019). Tending to destinations: Conceptualising tourism's transformative capacities. *Tourist Studies*, *19*(3), 279–294.

Krijnen, M. (2018). Beirut and the creation of the rent gap. *Urban Geography*, *39*, 1041–1059.

Kulusjärvi, O. (2020). Towards just production of tourism space via dialogical everyday politics in destination communities. *Environment and Planning C: Politics and Space*, *38*(4), 751–767.

Kurniadi, B. (2020). Defending the Sultan's land: Yogyakarta, aristocratic power and control over land in post-autocratic Indonesia. Doctoral Dissertation, Australia National University.

Li, T. M., and Semedi, P. (2021). *Plantation Life: Corporate Occupation in Indonesia's Oil Palm Zone*. Durham: Duke University Press.

Lund, C. (2023). An air of legality – legalization under conditions of rightlessness in Indonesia. *The Journal of Peasant Studies*, *50*(4), 1295–1316.

MacCannell, D. (1976). *The Tourist: A New Theory of The Leisure Class*. New York: Schocken Books.

Mosedale, J. (2006). Tourism commodity chains: Market entry and its effects on St Lucia. *Current Issues in Tourism*, 9(4–5), 436–458.

Peluso, N. L., & Vandergeest, P. (2020). Writing political forests. *Antipode*, *52*(4), 1083–1103.

Polanyi, K. ([1944] 2001). *The Great Transformation*. Boston: Beacon Press.

Pratt, S. (2022). Sustainable tourism development: Critically challenging some assumptions. *Tourism Planning & Development*, *19*(1), 26–36.

Ristiawan, R., Huijbens, E. & Peters, K. (2023a). Projecting development through tourism: Patrimonial governance in Indonesian geoparks. *Land*, *12*(1), 223.

Ristiawan, R., Huijbens, E. H. & Peters, K. (2023b). Apprehending land value through tourism in Indonesia: Commodification of rural landscapes through geoparks. *Tijdschrift voor economische en sociale geografie*, *115*(1), 170–186.

Slater, T. (2017). Planetary rent gaps. *Antipode*, *49*, 114–137.

Sloan, S., Campbell, M. J., Alamgir, M., Collier-Baker, E., Nowak, M. G., Usher, G. & Laurance, W. F. (2018). Infrastructure development and contested forest governance threaten the Leuser Ecosystem, Indonesia. *Land Use Policy*, *77*, 298–309.

Smith, E., Ristiawan, R. & Sudarmadi, T. (2022) Protection and repatriation of cultural heritage – country report: Indonesia. *Santander Art and Culture Law Review*, *8*, 383–406.

Smith, N. (1979). Toward a theory of gentrification: A back to the city movement by capital, not people. *Journal of the American Planning Association*, *45*, 538–548.

Smith, N. (1987). Gentrification and the rent gap. *Annals of the Association of American Geographers*, *77*, 462–465.

Tomassini, L., & Lamond, I. (2022). Rethinking the space of tourism, its power-geometries, and spatial justice. *Journal of Sustainable Tourism*, *31*(12), 2825–2838.

Young, M., & Markham, F. (2020). Tourism, capital, and the commodification of place. *Progress in Human Geography*, *44*(2), 276–296.

Yrigoy, I. (2019). Rent gap reloaded: Airbnb and the shift from residential to touristic rental housing in the Palma Old Quarter in Mallorca, Spain. *Urban Studies*, *56*(13), 2709–2726.

Yrigoy, I. (2023). Strengthening the political economy of tourism: Profits, rents and finance. *Tourism Geographies*, *25*(2–3), 405–424.

Part II

Cutting-Edge Tourism Interventions

6   Artistic Cartographies and Place-based Service Design for Making Places

Ella Björn and Satu Miettinen

Introduction

In tourism settings, Lapland is often marketed as a 'wilderness', which can give a false assumption of it being an 'empty' place and no one's land. Natural areas and places usually have deeply rooted cultural values and history, which might be invisible to newcomers (Li et al. 2023). Nature is a source of livelihood for many Sámi and Finnish people, including livelihoods like fishing, hunting, and reindeer herding, which should be considered when planning tourism services for these areas. In Sámi cultures, as in other indigenous cultures, nature and culture are not separate concepts but rather entangled. In indigenous knowledges people have always shared narratives and stories as ongoing processes (Silko 1981). Nature has usually been a significant factor in these narratives. The river can help to connect you to local stories and provide a sense of continuity and collectivity (see Kuokkanen 2007). Sápmi (the Sámi Homeland) covers the northern part of Finland, Sweden, Norway, and the Kola peninsula of Russia. On the Finland side, it covers the municipalities of Utsjoki, Inari, and Enontekiö. Utsjoki is located in the north-ernmost part of Finland near the Norwegian border, and most of the population is indigenous Sámi. Tourism offers sources of livelihood for many locals. However, problems have occurred in land use when reconciling local livelihoods, such as reindeer herding and tourism. Tourism in Utsjoki is mainly nature-based, and the main attractions are Arctic landscapes, the northern lights, silence, the closeness of Norway, and River Teno (Deatnu). Tourism has been very seasonal and focused mainly on salmon fishing in River Teno during summertime. The recent restrictions on salmon fishing have highlighted the importance of focusing more on year-round tourism to address seasonality issues. The municipality of Utsjoki aims to develop tourism in economically, culturally, and socially sustainable ways and based on the local people's needs and perspectives (Municipality of Utsjoki 2020). The role of non-human actors in tourism for preserving the quality of life of the locals and nature needs to be studied more.

Placemaking practices, where performativity and arts are utilised, can be used when preserving places with deep cultural and social meanings. New ideas for cultural tourism development can be raised when engaging stakeholders through these art-based practices such as creative and performative placemaking with

DOI: 10.4324/9781003449027-8

non-humans. Performative and embodied practices in the design process can help to gather tacit knowledge, which might otherwise be difficult to reach, giving a voice to the marginal groups that are not often heard. Performative and embodied practices outdoors also give agency to non-human elements in nature or human-built surroundings. Artistic placemaking supports the soul of the place, giving an understanding of deeply rooted cultural values in that place, which might be invisible to newcomers like tourists and visitors. Social interventions and innovations identified or established through creative placemaking in tourism can, at their best, offer ways for locals and tourists to feel connectivity. The connection to materials and place can be experienced through different cultural traditions such as traditional Sámi handicraft workshops, where the body can build a remarkable memory (Kugapi & Höckert 2022). They help break down dichotomies and cultural barriers by offering new opportunities for knowledge exchange. The ideal structure of the intervention is that it offers ecological, economic, and social benefits for the local communities.

This chapter studies the role of performative and art-based placemaking in building connections to places where rural and community-based tourism occurs. We envision that embodied and performative practices help to build respectful connections to other people, places, and nature in the co-creation processes for sustainable and place-based tourism solutions. The chapter is based on the collection of artistic processes carried out utilising photography, video, and performance as the means to design, test, and iterate new artistic ways and meanings for placemaking, as well as create artistic performances in natural environments. These experiments and interventions were conducted in two European Union (EU)-funded projects: 1) the Dialogues and Encounters in the Arctic (DEA) project 2020–2022, and 2) the Smart Cultural Tourism as a Driver of Sustainable Development of European Regions (SmartCulTour) project 2020–2023. The artistic processes involving embodied and performative practices were conducted in the municipalities of Inari, Enontekiö, and Utsjoki.

Theoretical Background

This chapter focuses on relational and pluriversal approaches to service design, placemaking, and performance in community-based tourism. Placemaking derives from the tradition of cultural geography (Lew 2017) associated with a sense of place (Othman et al. 2013), which is studied in this chapter through artistic and performative approaches. Placemaking can be either organic or planned, or a mix of both. Planned placemaking involves top-down efforts to design places and influence people's behaviours (Lew 2012), while organic placemaking is bottom-up, shaped by everyday life practices (Lems 2016; Lew 2017). The geographical notion of 'place' usually involves physical concepts such as topography, landscape, architecture, and the practices of the communities and their networks. With physical attributes not the only focus, different cultural practices, such as performing arts, cast place in a different light (Croose 2020). Richards (2017: 2) defined placemaking not just as a physical intervention but rather as a social practice that enables changes in thinking and doing. Placekeeping is another term for remembering the history of a place, maintaining the cultural and natural assets of the place, nourishing them, and developing respect (see Bedoya 2013; Moran 2016). Creative placemaking aims to recognise existing potential resources and engage the local community, which can be done by utilising arts and storytelling methods (see Lukić 2021; Markusen & Gadwa 2010; Richards 2017). Storytelling helps to reflect the role of non-humans in forming relations to places. The world is produced in diverse social and material relations where non-humans can be seen as actors (Law & Urry 2004; Ren 2011). Places and non-humans become 'alive' through stories, which can be demonstrated through embodied and performative practices. Storytelling fosters creativity within the participants of the collaborative works and enhances engagement and trust.

This chapter focuses on placemaking in the arctic natural environment or nature. It discusses the non-human agency in placemaking that is respected and recognised. Nature is recognised more as an author in scientific publications such as those focusing on Martuwarra River of Life (RiverOfLife et al. 2021) and Wann Country (Foster et al. 2020). Co-authorship can create new ways of looking at both ethical and copyright questions in arts and offer alternate views for this. In material-driven design, artists are experimenters, collectors, and natural agents in co-authoring material production that incorporates living non-human organisms and gives them agency. The exploration of contemporary material culture requires a border crossing between the artist's studio, laboratory, or residency and working between fields and disciplines at the intersection of art, science, and society (Berger et al. 2020). Ways to co-author are also discussed in modern cartography, which is studying new materialities and ways to present (Dora 2009).

Tourism has deeply rooted entanglements with colonial history, and telling stories can be a decolonising way to create new meaning-making and identity construction in tourism (Grimwood et al. 2019). Stories help share information, build connections and relationships, humanise people, and shape their sense of belonging (see Bruner 1987). Storytelling can be a decolonising action towards mutual respect, and it can be done through performance. Decolonising natural heritage explores multiple ways of interacting with the environment and non-humans, shaping environmental ethics similar to ethics of care (Boukhris 2020). The logic of gift is grounded in understanding the world, where relationships are extended to everyone and everything, emphasising reciprocity and responsibility towards others, including land (Kuokkanen, 2007: 7). As Kuokkanen (2007) described, the river can be seen as a gift and an element of transmotion, reminding people of the fluidity of its perceptions and interpretations. Genuine gift relationships with local community members and tourists are achieved when the joint interests surpass the economic ones, leading to positive behaviour and cultural experience (Assis & Peixoto 2019).

Creativity is seen as a tool to gain an advantage in regional development (Florida 2011). Placemaking interventions can offer ways to build social cohesion and understand the intangible cultural assets and heritage to build the destination's attractiveness (Lukić 2021). Placemaking activities combined with creative and

art-based methods may help the locals to express their emotions and feelings connected to place and the natural environment, which can be an important tool to
highlight the 'invisible' everyday life practices and cultural value of the place in
which non-human actors have a significant role, which is not often recognised. To
move towards more diverse and socially inclusive destination strategies, it is important to understand vernacular everyday life marginal practices, where creativity
occurs (Edensor et al. 2010). Everyday life practices of locals are recognised as increasingly important for tourists seeking authentic experiences. People look more
at the meaning they make with locals during their travels, and this meaning-making
between locals and tourists requires much creativity (Richards 2017). Performativity and creativity involve doing. When people do something with their bodies,
they focus on the moment, forgetting their cultural and demographic backgrounds,
which brings them closer to each other and makes them feel connected. Creative
tourism is a form of sharing experiences and feelings between tourists and locals
(Richards 2017) and a natural environment with non-human agents.

Placemaking relates to urban design and the inclusion of communities when designing, performing, or experimenting. It can also be used for experimenting with
co-authorship and performance in the natural environment (Miettinen et al. 2023).
Humanistic geographer Relph (1976) drew on the personal experience of an expe-
riencer who provided an analytical account of place identity. Place-based identity
unites a place's affective and cognitive aspects (Paasi 2003). Gender or social class
are substructures of self-identity. Artists and arts workers utilise creative placemaking and place-based arts at the centre of their communities. These communities
have evolved to become social, economic, and culturally empowering art forms.
Through the arts, artists with communities can accomplish more action than governments (Grodach 2011). 'Place ethic' is utilised in the context of place-based arts
with deep consideration, which demands consideration and respect for any place
that reaches further than merely drawing aesthetic inspiration derived from what is
known as the 'tourist gaze', which often results from a superficial and short-term
interaction with a place (Lippard 1997). Thus, place ethics considers the cultural
value of a place, acknowledging that it is deeply rooted in cultural meanings and
traditions that are often rendered invisible or silent.

The critical views on traditional placemaking approaches acknowledge the colonial conceptualisations of the 'site' (Foster et al. 2020), disregarding the agency
of nature in placemaking processes. Foster et al. (2020) explained the concept of
the country, including the ecologies of the living natural environment. The country
is a spiritual entity, not a separate institution of people. The places are not only
physical but always social as well, when formed through the movements along the
paths focusing on the condition of *being* in the world (Ingold 2011). This approach
gives agency to non-human and material actors in the process of making places.
Space is produced through embodied practices and experiences (Lefebvre 1991).
'Temporary' performances can help to create personal experiences in which community engagement can be built (Moran 2016). Movement, memory, time, and
expression can affect artists' creativity, attachment, and engagement with the landscape and place (Tilley & Cameron-Daum 2017). This chapter studies how creative

approaches can help evoke senses and emotions and how they can be transferred and utilised when preserving the sense of place where tourism occurs. It seeks to understand how designed and performative placemaking as interventions help build a feeling of belonging between people and nature; whether performative approaches in nature can offer a space where both visitors and locals can learn something about themselves, forget about their different positions, and feel connected to the time and place-sharing the experience; as well as the role of non-humans in building connections with others.

Research Methods

Our study focused on the considerations of place-based cultural and natural values, togetherness, and the role of non-humans studied through performative placemaking experiments in nature in Finnish Lapland. The study involved a placemaking approach to observe how deeply rooted cultural values and agency of non-humans can be utilised when preserving the sense of places and what kind of social benefits this could bring for community-based tourism. The values were studied through two case studies (DEA and SmartCulTour). In the first case study (DEA), performative approaches were more focused on gathering tacit and embodied knowledge of meanings, connections, and emotions related to place and belongingness. The second case study (SmartCulTour) studied the role of art-based placemaking in engaging stakeholders for community-based tourism development in different settings. Both case studies collected ideas on using body and feelings to engage participants in building connections in nature and developing new ideas on community-based tourism based on the place's deeply rooted and sometimes invisible cultural and personal values.

Methodologically, this study utilises arts-based research (ABR) approaches. At the core of ABR is its cyclicality. This is derived from an action research approach (Kemmis & McTaggart 2005) that is a long-term process in which each cycle operates independently but evolves based on the evaluation and results of previous cycles. New research knowledge is generated through artistic and practice-based case studies. Research design is an iterative research strategy based on learning from one's research process and findings (Thomas & Rothman 2013). As an outcome of the research design process, new theories, artefacts, and practices can be produced (Barab & Squire 2004). The research design of this paper includes several research cycles that are organised around ABR and designed around implementing art- and design-making processes with communities and non-humans.

Case Study 1: DEA – Dialogues and Encounters in the Arctic

DEA was a project funded by Interreg Nord and the Regional Council of Lapland to initiate dialogue between the indigenous peoples and researchers on challenges affecting the Arctic. The project aimed to raise possibilities for indigenous peoples to influence issues regarding their own culture. The DEA case study consisted of three performative experiments conducted between winter 2020 and winter 2022 in Finnish Lapland, in Sápmi. The participants of the experiments were designers and researchers working in the DEA and SmartCulTour projects. The first experiment occurred in January 2020 at Pulmankijärvi, a culturally important place for the Sámi people located in the municipality of Utsjoki on the Norwegian border. The second experiment was conducted at the Jäniskoski rapids and in a covered outdoor fireplace in Inari at the same time. Jäniskoski rapids are along the Juutua River and Juutua Trail. Juutua River is a home river for the Inari Sámi people, and it has been a place for gatherings and storytelling (Metsähallitus n.d.). The third experiment was enacted at Vuontisjärvi during autumn 2021 and winter 2022.

Placemaking in the natural environment was sensitive to co-authorship with the natural environment and recognised non-human agency. Development studies recognise decolonial discussions where indigenous ontologies, knowledge, and knowing are present, and collaboration is built on trust, reciprocity, relationships, and shared goals (Lloyd et al. 2012). Further, these priorities are actively shaped by families and non-humans. 'Humans, animals, plants, winds, rocks, spirits, songs, sunsets, and water, indeed all things, are connected in a web of kinship and responsibility' (Lloyd et al. 2012: 1076). In the placemaking experiments, we gave room for the affordances in the natural environment to create a deeply sensory space where we experienced not only the geohistorical but also a connection with a living ecosystem where both humans and non-humans belong and have agency. Placemaking was used to revitalise one's connection with the natural environment. Sometimes, there is a need to feel the connection through family-owned land. However, this is not always possible, as family grounds may be lost due to conflicts, changes in the borders of nation-states, or new policies. We experimented with how the natural environment can enable grounding through the familiarity of the location and create a healing or nurturing experience (see Figure 6.1). For example, bathing, floating, or swimming in moving water, such as a river, will co-author the experience and deeply impact one's embodiment and sensory place. All the locations we experimented with had a strong water connection.

This artistic placemaking aimed to explore the role of non-humans in interaction and temporality in performance and the environment. Artistic action can connect human bodies with the surrounding ecosystem and remind them of this connection, being embodied and part-of-nature creatures. The context of Sápmi was very specific and unique for the placemaking experiments. Before creative placemaking practices, one needs to understand the history of the place (Bedoya 2013). Much exploitation of the Sámi culture has occurred over the history of Finnish Lapland,

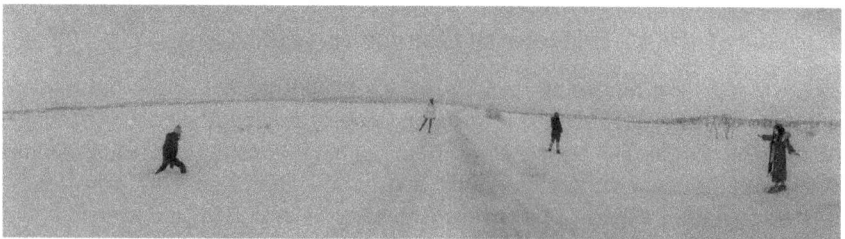

Figure 6.1 Performance through placemaking in Pulmankijärvi, Winter 2020

which has impacted the identity of the Sámi people and attitudes towards tourism. Performing in the indigenous lands calls for respect, humility, and sensitivity when creating placemaking experiments. Furthermore, it was discussed in some case studies that strong intergenerational connections with specific locations can carry immense meaning towards placemaking and connection with the place.

Case Study 2: SmartCulTour – Smart Cultural Tourism as a Driver of Sustainable Development of European Regions

The SmartCulTour project was funded by the EU as part of the Horizon 2020 framework supporting the regional development of sustainable cultural tourism. The project took a living lab approach, having six living labs in different geographical areas, including urban and rural destinations. Utsjoki was one of the rural destinations, and placemaking was tested as a place-based tool to support stakeholder engagement for cultural tourism planning. The experiments aimed to highlight the place-based cultural and natural assets and values that could benefit the local community when planning sustainable cultural tourism products and services. The placemaking experiment also covered placekeeping ideas on preserving tourism areas and how tourist behaviour can be guided towards mutual respect and understanding of nature. The case study comprised two performative placemaking experiments and one prototype feedback collection conducted in Utsjoki. Two different placemaking approaches to intervention were tested. The approaches were to design solutions for tourist behaviour through *designed placemaking* and to feel connectivity through *performative placemaking*. Both approaches utilised creativity and the active role of the visitor with non-humans in the natural environment.

The first performative experiment was conducted for the local tourism stakeholders in Utsjoki in February 2022. The experiment was done outdoors next to River Teno in one of the local tourism entrepreneur's premises in cold temperatures. The place was chosen because of its cultural and social importance to the locals, offering livelihood to many and being a central cultured landscape in Utsjoki. The artistic experiment had a performative approach involving both natural and designed affordances in play (see van Osch & Mendelson 2011). Designed affordances, such as handheld mini projectors, were used to project videos on the snow blanket next to a covered fireplace along the riverbank. The videos demonstrated different seasons in Finnish Lapland and showcased the meaning and fragility of

Figure 6.2 Performing placemaking in and near a covered fireplace in Utsjoki

the natural environment. One of the videos projected on the snow blanket utilised art-based ways to present and highlight the impacts of visitors' littering in nature, which was raised in earlier discussions with the local stakeholders. The participants were asked to look at the videos and reflect on how they felt in that current moment outdoors in the place and how they felt performing that feeling. After the performative part, the group went inside the covered fireplace to reflect on their feelings and share ideas about the experiment and place.

The second experiment occurred in May 2022, and the participants were locals and visitors of Utsjoki. The session had two parts. In the design part of the experiment, a sustainable solution to tourist experience and behaviour in nature was designed, prototyped, and tested in Village House Giisá in Utsjoki. The intervention, 'Traces in Utsjoki', was designed to tackle the issues relating to tourist littering and misbehaviour in nature. The intervention tested the role of designed affordances and tourist involvement in doing something good for the local community during their travels, such as collecting litter. The intervention included a photo gallery, a bingo game, and educative posters (see Figure 6.3). The idea of the photo gallery is that the visitor takes a photo of the found trace in nature with their mobile and uploads it to an open online photo gallery, which could be found on the municipality's webpage, or posts it on Instagram by using predetermined hashtags and the location where the photo was taken. This would make visible the spots where littering happens most often and offer ways to concentrate marketing efforts and guidance more on those sites where litter is found most. In addition, positive photos, such as photos of animal traces, can be spotted and posted to highlight the traces that belong in nature and have their own story and agency in the local culture. Children are activated through the playful part of the intervention, which is a bingo game for collecting litter and spotting different traces in nature. The informative part of the intervention included posters with guidance on how to act in nature in different situations, with the plan being to place posters in public places like the bathroom of the village house, which also serves as a visitor centre. Locals and visitors in Utsjoki familiarised themselves with the intervention and gave feedback and ideas on how the concept could be improved and benefit local people and nature.

The performative part of the placemaking experiment was held later the same day in the Ailigas Fell, a culturally important place for the locals. There are three

Figure 6.3 Collecting feedback on the Traces in Utsjoki intervention from local people and visitors in Utsjoki (SmartCulTour project)

Figure 6.4 Performative placemaking experiment in the Ailigas Fell (SmartCulTour project)

different Ailigas fells, which are all situated in the municipality of Utsjoki. All three Ailigas fells are natural sites, and our experiment occurred on the smallest Ailigas fell near the Utsjoki parish. The participants who took part in the experiment were visitors from Scheldeland (Belgium) and Huesca (Spain) living labs, including tourism stakeholders and tourism professionals. The experiment supported the findings of the first experiment involving a visitor perspective. The performative experiment deepened the idea of the Traces in Utsjoki intervention, bringing the bodily aspect into the study while studying the sensations and feelings of the participants in the fell and how the placemaking concept could be utilised in tourism. Five persons participated in the experiment. The participants were asked to choose a spot in a widespread area on top of the fell, listen to their bodies, senses, and emotions, and act upon them (see Figure 6.4). The experiences were discussed and reflected on later that day during the traditional Sámi dinner in a *lavvu* (traditional Sámi dwelling).

Findings and Discussion

The DEA case study focused strongly on performativity. The first experiment happened in Pulmankijärvi, and the landscape was misty and snow-covered. All one could see was whiteness, where the road and sky were blended into a beautiful scenery. The connection to places is formed because of the meaning created (Lew 2017), which was recognised during the study. Pulmankijärvi, Jäniskoski, and Vuontisjärvi all have meaning to local Finnish and Sámi people, and recognising that meaning might have affected participants meaning-making during the activity. Most of the participants were visiting these places for the first time. Before places can reach an identity of belonging, one must feel they *belong* and understand the social dynamics of the place (Bedoya 2013). Valuing these culturally important landscapes and showing them respect could be sensed from the participants' behaviour. The atmosphere of all three experiments was open and encouraging since the participants were already familiar with the art-based methods, which made it easier to get inspired and immerse oneself in the performing activity. Using the body as an instrument helps to evoke senses, and participants can let go of their worries and observe only their bodies while maintaining a sense of groundedness (Ludevig 2016). Nevertheless, the participants were able to let go and thrive in the moment;

their inner lives were still present, reflecting on how they behaved and performed. In the reflective discussion after the experiments, each participant pointed out everyday life happenings and concerns they have had in their personal lives. This built trust, empathy, and togetherness within the group, where nobody had known each other beforehand.

Indigenous knowledge systems often utilise storytelling methods. Stories helped the participants to make meaning and new connections, reflecting the experience to their personal lives. The performance was personal storytelling, and it evoked creativity; verbal stories helped to make these stories visible to others. One of the participants compared snow to sand, pondered how they are different materials, and pointed out her cultural background. Identity construction was visible in the participants' stories, where non-humans showed their agency in meaning-making processes. Snow was seen in many ways in the stories – as a deep barrier in which one had to wade to move further, or as endless white scenery to explore and see it all. Homesickness was also discussed with the participants. Re-emplacement is emotional work (McKay 2005), which was pointed out in the discussions of the participants' backgrounds and emplacement. Participants were newcomers to Finnish Lapland and the connection to their homes and different backgrounds were restored and elaborated on in the discussions. The SmartCulTour case study deepened the understanding of different place-related values in the natural environment and how they could be utilised in tourism development. There were three participants besides the research team. In the first outdoor experiment, the participants were too shy to perform. One reason for this might be that the participants did not have much experience in performing and art-based methods, which perhaps created barriers for them to improvise. Furthermore, there were new people in the group, and the temperature was low, which might have also affected the participants' performative mood. After a short while, one of the participants eventually immersed herself by lying down on a bench, putting her feet on the ceiling of a covered fireplace, and watching the video of Vuontisjärvi in autumn colours projected on the wall of the fireplace. The video on the ceiling allowed her to move on the bench. The projected videos raised interest among the participants, and during the reflective discussions, the different seasons and their importance were discussed. There is much cultural history in Teno River, and art-based methods could help to point out the importance of preserving these natural places. Additionally, videos could be utilised when streaming educative movies about the local culture and agency of nature for tourists during dark polar nights in the wintertime.

The second experiment of the case study gathered new ideas for the Traces in Utsjoki intervention. Although organic placemaking approaches support the soul of place more (Lew 2017), planned approaches like the Traces in Utsjoki intervention help manage organic placemaking by giving locals control over tourism development, which is the base of community-based tourism. According to the feedback received, Traces in Utsjoki was deemed rather an easy way to guide visitors in Utsjoki. The pictures of 'positive' traces in nature, such as animal prints that belong there, could improve information about the landscape and nature's part in local culture and identity. The performative and designed parts of the intervention supported each other by showing the role of non-humans in building connections and respect for places. In the performative part of the Ailigas Fell study, the visitors from Scheldeland and Huesca chose different spots to perform. One took shelter behind a rock, and the other one laid down on the ground. Most of the participants stayed still and did not move their bodies. One of the participants expressed later in the reflective discussion session that the place asked them to stay immobile and just feel the air on the face and breathe deeply. Participants had heard stories and the history of Sámi culture earlier on the same day, which made them reflect on what they had learned. During the reflective discussion, one of the visitors pointed out her concern about whether tourism is wanted in Utsjoki. She did not feel 'welcome' throughout her stay in Utsjoki and started to wonder whether the locals wanted tourism to be developed. Sometimes cultural differences can also affect how hospitality and welcome are shown and understood. Although enthusiasm and emotions might sometimes be lacking from the interaction, it might not necessarily mean that tourists are not welcome. New ideas utilising performativity in tourism were brought up in the reflective discussion. Posters could also use placemaking guidance for the tourists to conduct the bodily experiment in nature on their own and listen to their sensations while focusing on the moment and reflecting on the role of non-humans in the time and place.

Performative and design-driven placemaking interventions including embodied and artistic interactions create new affordances for performing in and learning about the natural environment. Performance through placemaking helps people utilise their embodied experience in the natural environment to prompt them to use their senses and emotions to improvise, discover, and strengthen the relationship between them and affordances in nature. Thus, humans design their experiences in nature through placemaking. The case studies showed that non-humans have an empowering role in building togetherness and evoking creativity. The performance helped the participants let go of all their unnecessary thoughts and focus on the moment, building new connections to each other and new memories of the place. Performative placemaking combined with designed and planned solutions, such as Traces in Utsjoki, works as an art-based placemaking act to preserve local natural places and allow visitors to feel connected and do good for the local community.

Acknowledgements

This article is based on two funded projects. The Dialogues and Encounters in the Arctic project is EU-funded by Interreg Nord and has received funding from Lapin Liitto (NYPS 20203486). The Smart Cultural Tourism as a Driver of Sustainable Development of European Regions project has received funding from the European Union's Horizon 2020 Research and Innovation Programme (Grant Agreement No 870708). We would like to thank all the collaborators and participants of both projects.

References

Assis, G. C. D., & Peixoto, R. C. D. (2019). Is tourism a gift? An 'ethnography of exchange' and the offer of so-called 'community-based tourism' experience in Anã/Santarém/Pará. *Revista Brasileira de Pesquisa em Turismo, 13*, 144–160.

Barab, S., & Squire, K. (2004). Design-based research: Putting a stake in the ground. *The Journal of the Learning Sciences, 13*(1), 1–14.

Bedoya, R. (2013). Placemaking and the politics of belonging and dis-belonging. *GIA Reader, 24*(1), 20–21.

Berger, E., Mäki-Reinikka, K., O'Reilly, K., Sederholm, H., & Schmidt, M. (2020). *Art as we don't know it*. Aalto University.

Boukhris, L. (2020). Decolonizing natural heritage: Knowledge, power and the political economy of tourism. In M. Gravari-Barbas (Ed.), *A research agenda for heritage tourism*. Edward Elgar, pp. 167–182.

Bruner, J. (1987). Life as narrative. *Social Research, 54*(1), 11–32.

Croose, J. (2020). Performing places: Carnival, culture and the performance of contested national identities. In T. Ashley & A. Weedon (Eds.), *Developing a sense of place: The role of the arts in regenerating communities*. UCL Press, pp. 139–161.

Dora, V. D. (2009). Performative atlases: Memory, materiality, and (co-)authorship. *Cartographica: The International Journal for Geographic Information and Geovisualization, 44*(4), 240–255.

Edensor, T., Leslie, D., Millington, S., & Rantisi, N. M. (Eds.) (2010). *Spaces of vernacular creativity: Rethinking the cultural economy*. Routledge.

Florida, R. (2011). 'The creative class': From the rise of the creative class: And how it's transforming work, leisure, community and everyday life (2002). *Canadian Public Policy, 29*(3).

Foster, S., Paterson, K. J., & Country, W. (2020). There's no place like (without) country. In D. Hes & C. Hernandez-Santin (Eds.), *Placemaking fundamentals for the built environment*. Palgrave Macmillan, pp. 63–82.

Grimwood, B. S., Stinson, M. J., & King, L. J. (2019). A decolonizing settler story. *Annals of Tourism Research, 79*, 102763.

Grodach, C. (2011). Art spaces in community and economic development: Connections to neighborhoods, artists, and the cultural economy. *Journal of Planning Education and Research, 31*(1), 74–85.

Ingold, T. (2011). *Being alive: Essays on movement, knowledge and description*. Routledge.

Kemmis, S., & McTaggart, R. (2005). Participatory action research: Communicative action and the public sphere. In N. K. Denzin & Y. S. Lincoln (Eds.), *The Sage handbook of qualitative research*. Sage Publications, pp. 559–603.

Kugapi, O., & Höckert, E. (2022). Affective entanglements with travelling mittens. *Tourism Geographies, 24*(2–3), 457–474.

Kuokkanen, R. (2007). *Reshaping the university: Responsibility, indigenous epistemes, and the logic of the gift*. UBC Press.

Law, J., & Urry, J. (2004). Enacting the social. *Economy and Society, 33*(3), 390–410.

Lefebvre, H. (2014). The production of space (1991). In J. J. Gieseking, W. Mangold, C. Katz, S. Low, & S. Saegert (Eds.), *The people, place, and space reader*. Routledge, pp. 289–293.

Lems, A. (2016). Placing displacement: Place-making in a world of movement. *Ethnos, 81*(2), 315–337.

Lew, A. A. (2012). Geography and the marketing of tourism destinations. In J. Wilson (Ed.), *The Routledge handbook of tourism geographies*. Routledge, pp. 181–186.

Lew, A. A. (2017). Tourism planning and place making: Place-making or placemaking? *Tourism Geographies*, *19*(3), 448–466.

Li, H., Lüthje, M., Björn, E., & Miettinen, S. (2023). Fostering stakeholder engagement in sustainable cultural tourism development in nature-based sites: A case study on using methodological layering of art-based methods. In A. Mandić & K.W. Sandeep (Eds.), *The Routledge handbook of nature-based tourism development*. Routledge, pp. 183–200.

Lippard, L. R. (1997). *The lure of the local: Senses of place in a multicentered society*. New Press.

Lloyd, K., Wright, S., Suchet-Pearson, S., Burarrwanga, L., & Country, B. (2012). Reframing development through collaboration: Towards a relational ontology of connection in Bawaka, North East Arnhem Land. *Third World Quarterly*, *33*(6), 1075–1094.

Ludevig, D. (2016). Using embodied knowledge to unlock innovation, creativity, and intelligence in businesses. *Organizational Aesthetics*, *5*(1), 150–166.

Lukić, I. V. (2021). Placemaking, local community and tourism. *Croatian Geographical Bulletin*, *83*(1), 77–104.

Markusen, A., & Gadwa, A. (2010). *Creative placemaking*. National Endowment for the Arts.

McKay, D. (2005). Migration and the sensuous geographies of re-emplacement in the Philippines. *Journal of Intercultural Studies*, *26*(1–2), 75–91.

Metsähallitus. (n.d.). Juutuan luonto- ja kulttuuripolku. Retrieved 11 August 2023 from: https://www.luontoon.fi/inari/reitit/juutuanluontopolku.

Miettinen, S., Björn, E., & Alhonsuo, M. (2023). Place-based service design through placemaking and performance. In *ServDes 2023 Conference: Entanglements and flows: service encounters and meanings*, Rio de Janeiro, Brazil.

Moran, J. (2016). The role of performing arts in place. In J. Schupbach & D. Ball (Eds.), *How to do creative placemaking. An action-oriented guide to arts in community development*. National Endowment for the Arts, pp. 28–31.

Municipality of Utsjoki (2020). Utsjoen matkailun kehittämisohjelma ja maankäyttösuunnitelma. Retrieved 23 August 2023 from: https://www.utsjoki.fi/wp-content/uploads/2022/01/utsjoenmatkailunkehittmisohjelmajamaankyttsuunnitelma.pdf.

Othman, S., Nishimura, Y., & Kubota, A. (2013). Memory association in place making: A review. *Procedia—Social and Behavioral Sciences*, *85*, 554–563.

Paasi, A. (2003). Region and place: Regional identity in question. *Progress in Human Geography*, *27*(4), 475–485.

Richards, G. (2017). Making places through creative tourism. In *Proceedings of the International Conference on Culture, Sustainability, and Place: Innovative approaches for tourism development*, Ponta Delgada, Azores, Portugal.

RiverOfLife, M., Taylor, K. S., & Poelina, A. (2021). Living waters, law first: Nyikina and Mangala water governance in the Kimberley, Western Australia. *Australasian Journal of Water Resources*, *25*(1), 40–56.

Relph, E. (1976). *Place and placelessness*. Pion Limited.

Ren, C. (2011). Non-human agency, radical ontology and tourism realities. *Annals of Tourism Research*, *38*(3), 858–881.

Silko, L. M. (1981). Language and literature from a Pueblo Indian perspective. In L.A. Friedler & H. A. Baker (Eds.), *English literature: Opening up the canon. Selected papers from the English Institute, 1979*. The Johns Hopkins University Press, pp. 54–72.

Thomas, E. J., & Rothman, J. (2013). *Intervention research: Design and development for human service*. Routledge.

Tilley, C., & Cameron-Daum, K. (2017). Art in and from the landscape. In C. Tilley & K. Cameron-Daum (Eds.), *An anthropology of landscape: The extraordinary in the ordinary*. UCL Press, p. 346.

van Osch, W., & Mendelson, O. (2011) A typology of affordances: Untangling sociomaterial interactions through video analysis. In *Proceedings of the 32nd International Conference on Information Systems*, Shanghai, China.

7   The Tension is Rising

Storytelling as an Intervention

Ondrej Mitas, Juriaan van Waalwijk, Bertine Bargeman, Licia Calvi, Lotte van Esch, Moniek Hover, Wilco Boode, and Marcel Bastiaansen

Introduction

Competition in destination marketing and management has become increasingly intense and complex. Numerous locations are being developed and communicated as potential tourism destinations. At the same time, local political shifts in heavily visited cities and regions now place new demands on destination management organizations (DMOs) (Richards & Duif, 2018; Volgger & Pechlaner, 2014). DMOs are asked to not only continue to attract tourists, but to steer them toward locations and behaviors that optimize tourism impacts on local residents (Eckert et al., 2019). This trend clearly affects tourist attractions such as theme parks, cultural landscapes, transportation hubs, and museums. All of these facilities are now expected to provide quality experiences to tourists, who should then produce positive word-of-mouth for the destination; and to local residents, who should then increasingly identify and take pride in the place or city they live (Richards, 2020). Thus, there is substantial pressure to identify and implement interventions that provide appealing, engaging experiences to tourists and local residents alike. The present chapter examines the potential of *storytelling* as an intervention to improve an experience available to both tourists and local residents, namely visiting a city museum.

Evidence of people telling each other stories extends back as far as archaeological records allow. Academic literature has recognized storytelling as one of the oldest and most widely recognizable methods humans have developed for communicating emotionally laden information with one another (Baker, 2018; Fu et al., 2023). As emotions are a crucial building block of experiences (Bastiaansen et al., 2019), implementing storytelling is one of the most promising potential interventions to improve the value of tourist attractions to both visitors and residents (Calvi & Hover, 2021; Green, 2021). It is important to state upfront that we define storytelling narrowly in this chapter as "the creative process (as situated in a context) … of inventing, (re)writing, designing, and/or staging of stories with implicit and/or explicit expressions (or 'texts') as a result" (Hover, 2013, p. 26). We do *not* include mere rhetorical improvements in marketing materials under "storytelling". In contrast to marketing techniques, we conceptualize storytelling in this chapter as a method for designing the experience itself, not merely communicating it.

DOI: 10.4324/9781003449027-9

With that in mind, we describe and evaluate the implementation of storytelling at a city museum set in a medieval palace, the Markiezenhof, in the small Dutch city of Bergen op Zoom. We first review the role of storytelling in communicating and triggering emotion, then explain the role of emotion in (visitor) experiences. Finally, we explain the empirical findings from a field study which measures emotion experienced by Markiezenhof visitors. These findings empower us to make conclusions about the effectiveness of a specific storytelling intervention at the level of single museum. We conclude by considering how such storytelling interventions can contribute to experience design needs on a regional level.

Literature Review

Storytelling

In line with Bal and Van Boheemen (2009), we define stories as narrative sequences of (chrono)logically connected events that are caused by or happen to certain characters. Characters may be human, animal, or even object. Attempts to explain differences between stories and other communication forms, often with the goal of creating a "formula" or model by which new stories can be reliably constructed, date back thousands of years. McKee (2010) mentions that Aristotle was the first to write about the underlying structure of stories. He recognized a "three act" structure: beginning, middle, and end. Within this structure, Artistotle's advice was to write stories about characters who have something to lose. A story should begin "in medias res" – in the midst of things – that is, at a specific point of the protagonist's life. Then, something happens that sets the events in motion and calls the protagonist to act. This is called the motoric moment (Hover, 2013), inciting incident (McKee, 2010) or call to adventure (Campbell & Moyers, 2011). When the audience experiences this inciting incident, a so-called "major dramatic question" is sparked in their mind. They subconsciously ask themselves "how will this turn out?" (McKee, 2010), thus raising the emotional tension through narrative turning points towards the climax. At the climax, where the audience's emotional reactions to the story are thought to be the most intense, an answer to the major dramatic question begins to unfold. The resulting resolution is said to cause a final, definitive decline in emotional intensity. Climaxes often revolve around a major reversal in the protagonist's life. A story can then end on either a positive or a negative charge, depending on the genre (i.e., fairytale vs. tragedy). Goldman argues that "the key to all story endings is to give the audience what it wants, but not the way it expects" (McKee, 2010, p. 433).

Another aspect of storytelling which is believed to evoke emotional involvement in audiences is empathy (Fu et al., 2023); McKee and Gerace (2018) explain empathy as a combination of identification and insight. When something happens to the protagonist in a story, the story audience experiences it as happening to them personally. Gordon et al. (2018) claim that empathy is a crucial component for bringing about narrative transportation – that is, the feeling of being drawn into the story.

Two conclusions about the effectiveness of stories are clear from this literature. First, empathy with characters is crucial to triggering emotions in a story's audience. Second, the tension experienced by characters rises throughout at least the first half of a story – often nearly to the story's end – allegedly triggering rising levels of emotional arousal in a story's audience. It is this second conclusion we use in the present chapter to assess the effectiveness of a storytelling intervention.

Storytelling is used as an intervention at destinations precisely to facilitate meaningful experiences by evoking emotional arousal in tourists visiting that destination, and at museums in particular (Skolnick, 2005). By associating content and information related to a location with emotional responses within this experience, tourists may better remember the place (Bastiaansen et al., 2019) and develop a positive emotional connection to it (San Martín & Del Bosque, 2008). As a matter of fact, storytelling is increasingly used as a tool for placemaking, i.e., in designing experiences for visitors at tourist destinations (Calvi & Hover, 2021) and in heritage tourism, for instance as a tool for exhibition design in museums (Calvi et al., in press). In tourism practice, storytelling is used beyond mere destination marketing, but also to stage experiences for visitors and by visitors themselves to co-create their own experiences, by sharing their personal stories of their visit. To implement storytelling in a museum context, designers make use of appealing artifacts, descriptive texts, and other information sources (e.g., digital and/or interactive) to stimulate visitors' imaginations and act as vehicles of emotion and subsequent memory (Chronis, 2012). Positive and meaningful memories could lead to both word-of-mouth recommendation and repeat visits which enlarge their personal "storybook" about a museum (Bargeman et al., 2022). Previous research indicated that a consistent approach to application of storytelling to the field of tourism is still lacking (Baker et al., 2019). Thus, it is difficult to conclude whether storytelling interventions at tourist destinations are inherently effective.

Nevertheless, research in the field of museum studies shows that storytelling is effective in eliciting both positive and negative emotions in museum visitors and in facilitating meaning making for them (Jelincic, 2020) and that this might even lead to transformational experiences and repeat visits. Unfortunately, these findings are based on self-report and thus subject to recall biases (Bastiaansen et al., 2019). Because these biases warp and collapse temporal dynamics of experiences (Strijbosch et al., 2021b), it is impossible to understand if emotions triggered by stories in museum contexts follow intentioned models of storytelling, or create an entirely different and unintended emotional profile. To appreciate the importance of this distinction, it is necessary to first clarify the role of emotion in experience.

Emotion in Experience

The power of stories to intervene in how destinations, and museums in particular, are experienced, remembered, and visited allegedly arises from their ability to trigger emotions (Gabriel, 2000; Mitas et al., 2020a; Mitas et al., 2020b). To understand the psychology of this process, it is necessary to examine the role of emotions in the workings of the mind. Bastiaansen and colleagues (2019) conclude that emotions are the key by which experiential stimuli, such as reading, watching, or listening to a story, become encoded in memory and trigger behavior. More specifically, the neural processes of memory become activated when emotional arousal – the extent to which an emotion is consciously intense – reaches above a certain minimum threshold level. Thus, it is the emotional intensity of stories that allows an audience to remember them and pass them on. As discussed earlier, a defining characteristic of most storytelling forms is escalating emotional intensity. As such, the form of stories makes the facts contained therein memorable *by design* and stimulates narrative enrichment and imagination (Chronis, 2012).

Beyond engaging memory making, emotions are crucial, albeit indirect, drivers of behavior (Baumeister et al., 2007). Emotions serve as information, thus directing the possible actions an individual is considering at any one moment (Fredrickson, 1998). As destination management interventions seek to trigger either directly emotional goals (e.g., feeling pride in one's city) or, more often, behavioral ones (e.g., sharing or recommending a destination, as in destination branding; Bastiaansen et al., 2018), it follows that triggering emotions is a component of nearly all successful destination management interventions. Even a rather cut-and-dried intervention, such as closing a street to traffic, is mediated by emotions: would-be drivers see barricades and realize the only emotionally acceptable option is to turn around. This process is pre-conscious, but underlies all decision making (Schwarz, 2000). Thus, whether aimed at tourists, residents, business operators, or all three, destination interventions with reliable emotional consequences, such as pronounced positive emotions and behavioral intentions, are of great interest for destination management organizations.

Need for Empirical Validation

While the power of stories to trigger emotion is anecdotally well-accepted, and the role of emotion in experiences clearly demonstrated, there is a gap in empirical research on the use of storytelling in experience design and its effect on emotion (Nabi & Green, 2015). While the established story structure – with rising tension – is often used in experience design and is a widely applied tool in designing museum exhibits (Bal, 2007; Calvi et al., in press; Francis, 2020; Macleod et al., 2018). Whether this design actually triggers progressively rising levels of emotional arousal in visitors is not yet researched. Furthermore, it is also not well understood if the progressive increase in emotional arousal actually makes for a better experience (Mitas et al., 2020b). In the present study, we test this hypothesis by comparing skin conductance responses, an objectively measured physiological marker of emotional arousal, across four rooms in the Markiezenhof palace in which a story unfolds.

Context: Markiezenhof

The Markiezenhof ("Marquise Palace") is a city palace in Bergen op Zoom, a town in the south of the Netherlands. Since the 1980s, it has hosted a museum about the history of the city. To enhance and specify the city's destination brand, an intervention was carried out to redesign this museum around the concept of secrets. More specifically, four period rooms with 17th and 18th century-interiors were redesigned using the true story of the marquise Marie Anne van Arenberg (1689–1736). The resulting exhibition is called "The Marquise's Secrets". The redesign was carried out by the authors of the present chapter, and technically implemented by a commercial firm with experience in museum exhibit design and construction. Apart from building a storyline with emotional tension and a clear climax, we also aimed to trigger empathy among visitors for the protagonist. The story's events are told or shown by using physical objects and digital techniques (Bargeman et al., 2022).

Marie Anne's life was filled with secrets and drama. At a very young age, she was married to the much older Marquis François Egon. A daughter was born, but François passed away just three years later. Marie Anne started an affair with a sta-blemaster, Simon de Maisy, and later married him in secret. A conniving cardinal threatened to bring this secret to light, after which Marie Anne's mother forced her daughter to choose between her husband and her daughter.

The four rooms show visitors dramatic events in Marie Anne's life. We reframed the first room as a "Wedding Room". Here visitors are guests at Marie Anne's wedding to François Egon. The second room, a "Chamber of Secrets", features a fancy canopy bed and a dressing curtain. Digital projections are used to create a shadow play on the dressing curtain. The shadow play portrays several turning points in Marie Anne's life: the birth of her daughter, the death of her first husband, and her affair with the stablemaster. A talking painting features the conniving cardinal who tells us about the secret wedding between Marie Anne and Simon, and how he intends to use this against her in order to obtain more power in the region. The third room is a "Room of Choice", centered around a dining table at which visitors are clearly asked to sit. Once sensors detect that visitors are sitting, audio-visual effects start. Herein visitors hear the fierce argument taking place between Marie Anne and her mother. Marie Anne's young child expresses her distress. Marie Anne is forced by her mother to choose between giving up her husband or giving up her child. Visitors are then asked to take the role of Marie Anne and, in front of an interactive family portrait, sign either the divorce papers or the custody papers, after which one or more characters disappears from the portrait. The last room presents "The End". Here we read how the story ended for Marie Anne and the other characters. She ultimately chose to give up custody of her child. She was forced out of the Markiezenhof with Simon and died quite young, leaving three more children.

In short, the first room sets up the context of the story and introduces the characters. The motoric moment is the wedding of young Marie Anne to the much older François. The second room features several turning points (birth of child, death of husband, and start of affair). The third room works towards the climax (having to make an impossible choice) and the final room offers a resolution. Thus, the rooms divide the established structure of stories into several stages with rising tension. The goal of the present study was to determine if visitors to The Marquise's Secrets exhibition experienced escalating emotional arousal across the four rooms. In other words, we aimed to test the effectiveness of the exhibition in triggering a pattern of emotional responses consistent with the intention behind the story being told.

Methods

It would be possible to measure emotional reactions to an exhibition by asking participants questions repeatedly during the experience (Kono et al., 2022), following them as participant observers (Bargeman et al., 2022; Calvi et al., in press; Mitas et al., 2012), or asking them to reconstruct the experience scene-by-scene afterward (Strijbosch et al., 2019). However, all these methods are subject to response or recall biases, which can be very substantial in emotion variables (Bastiaansen et al., 2019; Gilbert, 2006; Wirtz et al., 2003; Zajchowski et al., 2016). Thus, following the recommendation of Bastiaansen et al. (2019), we measured psychophysiological arousal, specifically skin conductance, combined with location tracking, to measure change in emotional arousal over the course of the exhibition. This approach has been successfully used before in museum exhibits (Kirchberg & Tröndle, 2015; Mitas et al., 2020a) and allows visitors to explore the exhibition with their chosen companions, at their own pace, while their emotional arousal is unobtrusively and continuously recorded in real time. We used questionnaires before and after the visit to record participant characteristics and experience evaluations.

Sample

Participants included in the study were visitors of the museum intercepted prior to entering the exhibit. They had to be 18 years old and to agree to partake in the research. If they consented to participate, they were taken to a separate room near to the exhibition where they received further instructions, filled in a pre-questionnaire, and were equipped with wristband wearable devices and smartphones for tracking skin conductance and location, respectively. They then were brought back to the exhibit where they could look around at their own pace. After the exhibit participants were picked up and brought back to the briefing room to fill out the post-questionnaire and give back the devices. To thank the respondent for their participation, they received a coupon for a free cup of coffee and dessert. The final sample comprised 84 participants.

Measures

Experience Evaluation

To measure participants' judgment of their experience of the Markiezenhof as a whole, we used a 0–10 (not at all likely – extremely likely) single-item measure of intent to recommend (Reichheld, 2003). This item is considered a widely used standard for evaluating the quality of experiences both in the tourism industry, and in academic studies of tourism experiences. A limitation of this single-item measure is that it does not capture the multidimensional structure of complex experience evaluation judgments. We chose to make this trade-off to keep participant burden to a minimum while including both continuous experience measurement and evaluation of the museum visit as a whole.

Emotional Arousal

Following recent precedents in continuous emotional arousal measurement in museums (Mitas et al., 2020a; Mitas et al., 2020b), we used the Empatica E4 wearable wristband to measure skin conductance as a proxy for emotional arousal. While this device measures skin conductance using electrodes on the underside of the wrist by default, we chose to replace these with leads to single-use electrodes attached to the underside of the medial joints of the D2 and D3 fingers of the non-dominant hand (Figure 7.1). Using electrodes on the fingers as opposed to the wrist reflects current best practice for mobile skin conductance measurement (Li et al., 2022; Strijbosch et al., 2021a).

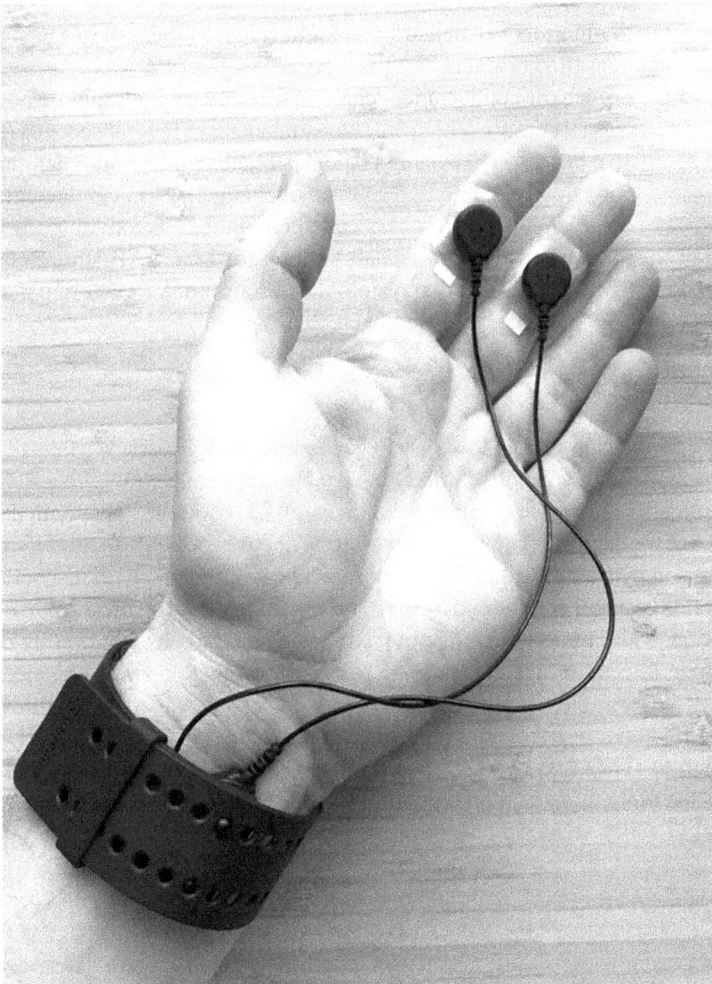

Figure 7.1 Placement of Empatica E4 wristband for skin conductance recording

The Empatica E4 records the raw conductance between two electrodes at a frequency of 4 Hz. The device also records timestamps at the moment its button is briefly pressed. Using button presses at the beginning and end of each participant's visit meant that data from before and after the visit were cut and deleted. Subsequently, the data were scanned and cleaned from motion artifacts. When substantial movements of the hand or arm alter the pressure of electrodes against the skin, skin conductance changes in a way that is unrelated to emotional arousal. These so-called motion artifacts in the data are characterized by rapid spikes or troughs whose origin cannot plausibly be physiological. In accordance with current practice, we used the ArtifactZ tool for MATLAB to remove motion artifacts. ArtifactZ implements a 20-second sliding window to identify peaks or troughs greater than three standard deviations. We visually inspected these suspected artifacts, and if their shape or corresponding accelerometer data suggested they were due to movement rather than emotion, we deleted them and replaced the gap in the signal by linear interpolation (Strijbosch et al., 2021a).

Furthermore, the skin conductance signal is affected by the rapid responses of the body to emotional stimuli, which emerge and fade over several seconds (Braithwaite et al., 2015), but also changes in response to ambient and body temperature, which unfold more slowly. These two sources of change in skin conductance must be statistically separated using a process called continuous deconvolution. We implemented this process using the Ledalab toolbox for MATLAB and retained the phasic component, which reflects physiological arousal due to emotional stimuli, for further analysis (Benedek & Kaernbach, 2010).

Location

To record participant location, we asked each participant to carry a Motorola Moto E5 smartphone with an application to receive beacon signals throughout their visit. The application recorded signal strength from 25 Estimote Bluetooth beacons which were mounted throughout the exhibit. While it is possible to identify participant location by triangulating the three nearest beacon signals with respect to their locations, this signal is inherently noisy and unreliable due to signal reflection and obstruction. More importantly, due to the open layout and close proximity of story elements in the exhibition, it is possible for participants to be near one element of the story (for example, the display of divorce and custody papers in the third room) while being actively engaged with a different element (such as the conflict at the dinner table). Thus, we chose to use the nearest beacon to identify which room participants were in, at a frequency of once per second (1 Hz). This allowed us to ascertain to which step of the story participants were being exposed.

Data Analysis

We first generated descriptive statistics of participant characteristics and experience evaluations in terms of frequencies for discrete variables, and mean, standard deviation, skew, and kurtosis for continuous and Likert-type variables. Then, focusing on skin conductance, we used the lmer() multilevel implementation of repeated measures ANOVA in R 4.2.2 to model z-standardized phasic skin conductance response as a function of which room each data point originated from. Maximum likelihood estimation was used. Follow-up pairwise comparisons were conducted by re-running the model using the first, second, and third room respectively as reference categories.

Findings

Participants

Our sample comprised slightly more women (51.8%) than men, with no participants identifying as non-binary. The sample was relatively well-educated, with 39.8% of participants possessing a bachelor's degree, while 27.7% reported a postgraduate degree. Age of participants was broadly distributed around a mean of 47 years (sd = 18 years) and ranged from 18 to 77 years.

Experiences of Visiting the Exhibit

Participants spent an average of 21.9 minutes visiting the exhibition (sd = 7.8 minutes). The third room, showing the story's dramatic climax with multiple interactive elements, featured the longest mean visit duration (6.6 minutes; sd = 1.4 minutes), closely followed by room 1 (mean duration = 6.4 minutes; sd = 2.6 minutes) and room 2 (mean duration = 5.6 minutes; sd = 2.3 minutes). The average visit to the last room, which resolved and concluded the story, was substantially shorter at 3.4 minutes (sd = 2.6 minutes; Figure 7.2).

As is common for tourism experiences, intent to recommend the Markiezenhof was slightly negatively skewed, with a mean of 7.18 (sd = 1.92). There were no significant correlations between duration in any of the four rooms of the exhibit and intent to recommend (all Pearson's r's < 0.2; all p's > 0.1). Intent to recommend also did not differ by age, gender, or education level (all p's > 0.1).

Emotional Arousal Over the Course of the Story

The mixed-effects implementation of repeated ANOVA with skin conductance response as a function of room from which each data point originated, nested within participants, fit the data significantly better than a model with no predictors (Chi-square -2LL (4) = 23219, p < 0.001; Table 7.1). All pairwise differences between rooms were significant at the p < 001 level even with the conservative Bonferroni correction (all T-values of pairwise comparisons between rooms were 24.29 (44620) or higher). Figure 7.3, showing the average skin conductance response per room, indicates that each successive room was more emotionally arousing than the previous one, with the last room – where visitors learn how the story ended – clearly being the most intense.

Based on the skin conductance apparently increasing in a near-linear fashion across the four rooms, we conducted several follow-up analyses to probe this effect. First, we merged questionnaire data with skin conductance and location data. We created a model where a predictor term represented linear growth across the

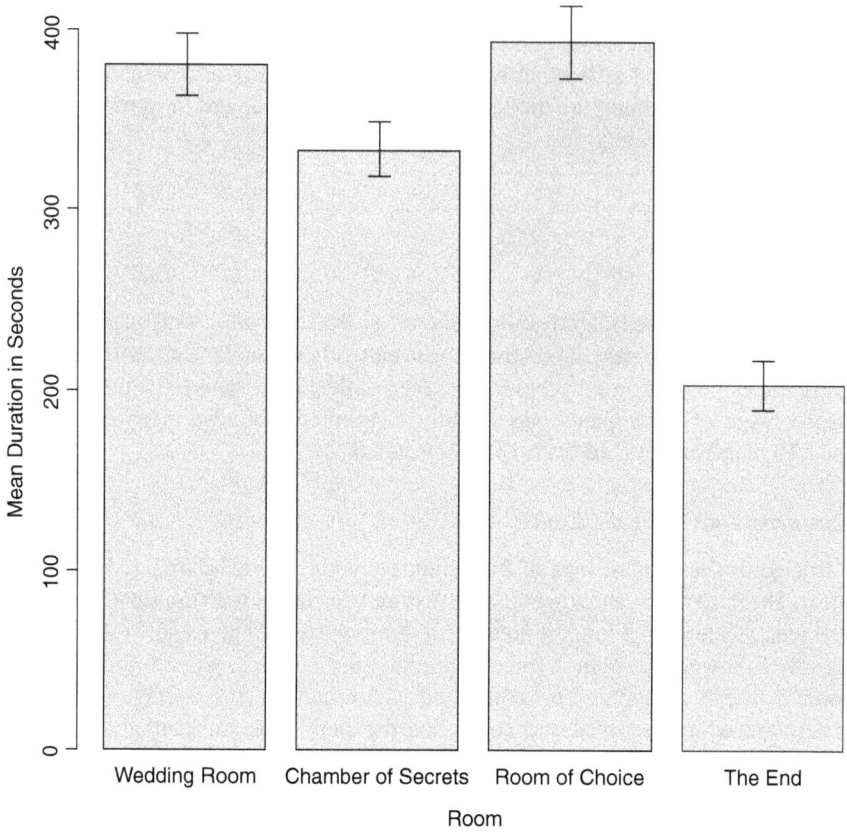

Figure 7.2 Average duration of visit by room of "The Marquise's Secrets"

Table 7.1 Mixed-effects model of skin conductance as a function of room

Outcome	Predictor	Coefficient estimate	SE	T	Model AIC
Phasic skin conductance					-294084
	(Intercept)	0.1292	0.0094	13.76***	
	Wedding Room vs Chamber of Secrets	0.0173	0.0007	24.29***	
	Wedding Room vs Room of Choice	0.0671	0.0007	98.05***	
	Wedding Room vs The End	0.1127	0.0008	134.96***	

Note: ***=p<0.001 for Satterthwaite approximation of T-test comparing parameter to 0.

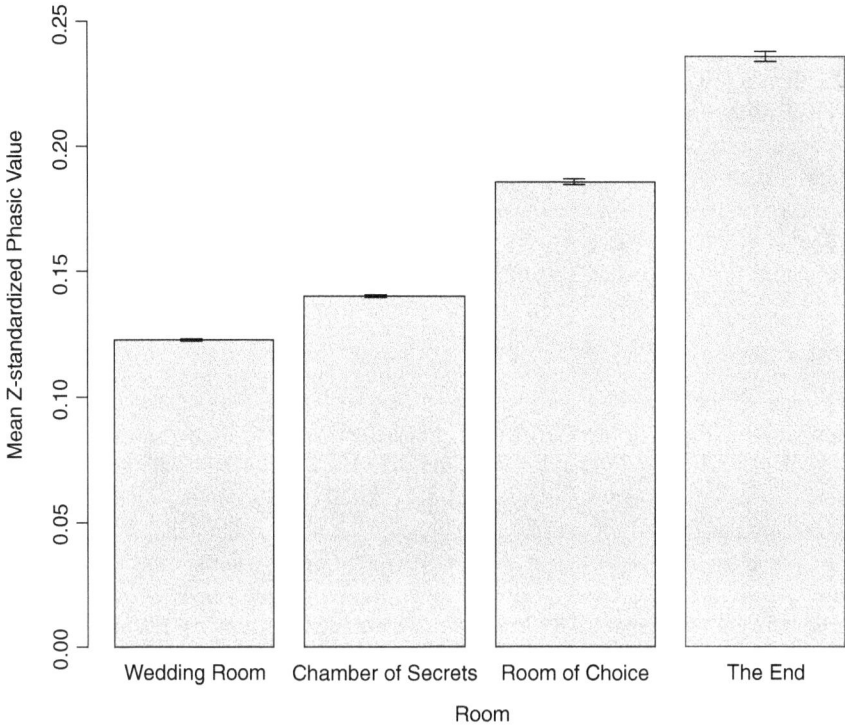

Figure 7.3 Average skin conductance per room of "The Marquise's Secrets"

Table 7.2 Mixed-effects models of linear growth in skin conductance over the course of "The Marquise's Secrets"

Outcome	Predictor	Coefficient estimate	SE	T	Model AIC
Phasic skin conductance					-293131.5
	(Intercept)	0.0842	0.0094	8.956***	
	Linear growth	0.0375	0.0003	149.474***	
Phasic skin conductance					-293149.5
	(Intercept)	0.0955	0.0109	8.785 ***	
	Linear growth	0.032	0.0012	26.853***	
	Intent to recommend	-0.0015	0.0007	-2.039*	
	Linear growth X Intent to recommend	0.0007	0.0002	4.720***	

Note: Signif. codes: 0 "***"; 0.001 "**"; 0.01 "*" 0.05; "."0.1; "" 1

four rooms (room 1 = 1, room 2 = 2, etc.). As expected, this linear growth term significantly predicted skin conductance response (T = 149.474 (44620); p < 0.001). We then allowed this linear growth term to interact with intent to recommend. The resulting interaction was positive and significant (T = 4.72 (44620); p < 0.001). In other words, high-recommending participants experienced a stronger linear increase in emotional arousal from one room to the next. Notably, when modelled for intent to recommend of 6, 8, and 10 on an 11-point scale, it was shown that the effect of moving from room to room on emotional arousal is dominant, and including intent to recommend as an interaction term influences this effect only slightly.

Discussion

The goal of the present chapter was to assess the effectiveness of storytelling as an intervention. Previous literature asserted that stories tend to follow a consistent pattern of increasing emotional tension, and that this translates into increasing emotional arousal in a story's audience (McKee, 2010). Because emotional arousal is understood as crucial for memory and behavior, it was assumed that telling stories associated with a destination would improve that destination's impact on visitors and lead to desirable behavioral intentions. In theory, such an intervention could promise to not only make a destination more attractive to tourists, but also increase pride or sense of identity among local inhabitants (Chronis, 2012). We tested this theory at "The Marquise's Secrets", an exhibition at the Markiezenhof city palace and museum in Bergen op Zoom, the Netherlands. We specifically aimed to understand if emotional responses to this exhibition followed established models of storytelling, wherein increasing tension among characters in a story triggers increasing emotional arousal (McKee, 2010).

Our data clearly showed an increase in emotional arousal from the beginning of the story to the end within the four rooms where the exhibition took place. This increase was linear, and slightly stronger for participants who left the exhibition more likely to recommend it. Thus, we conclude that the implementation of storytelling in "The Marquise's Secrets" is effective in triggering escalating emotional arousal, and individuals who experienced more such emotional escalation were also more likely to recommend the exhibit.

The data do show an interesting digression from established models of storytelling, however. While endings allegedly feature a sense of emotional relief that comes from resolving the tension-causing conflicts between characters, the last room in "The Marquise's Secrets" featured unambiguously higher emotional arousal. A possible explanation is that visitors positively reflect on the exhibition as a whole in the last room, a phenomenon known as savoring (Bryant & Veroff, 2017). The potential for savoring is accentuated in the last room via a photo booth, which encourages participants to create and share a socially relevant memory of their visit. The use of photography as a savoring mechanism has been empirically validated (Kurtz & Lyubomirsky, 2013).

Another possible explanation is found in the details of how storytelling is implemented in this exhibition. There is a slight difference between the climax in the *interactive experience design* and the climax of the *story*. In the third room the visitors can step into Marie Anne's shoes and sign either the divorce papers or the custody papers, after which the interactive family portrait reacts. But the true climax of the *story* is what Marie Anne herself really did, and this is disclosed only in the fourth room. Here we can also read the resolution, i.e., how it all ended for Marie Anne and her children. A previous qualitative study of the "The Marquise's Secrets" showed that visitors not only felt empathy for Marie Anne, but also for her young child (Bargeman et al., 2022). Reading that Marie Anne actually left her first-born child behind can lead to intense emotional arousal.

While this resolution may be in line with how resolutions are typically conceptualized in storytelling models, they may be associated with higher rather than lower emotional arousal in participants (Petrovic et al., 2022). This possibility, and disambiguating it from savoring, merit further investigation in more tightly controlled storytelling experiences.

Conclusion

The present chapter shows that storytelling can be implemented at tourist attractions with effective emotional impact on visitors. Such visitors may be tourists, or local residents aiming to "discover" their own city. The findings raise numerous new questions about the possible implementations of storytelling at both the attraction (e.g., museum) level and at the regional level in destination management. How could such effective storytelling implementations be communicated in (marketing) instruments for DMOs in cities and tourist destinations, not only to attract visitors, residents, and other interested groups to a museum or city, but also to manage flows of people, to reduce overtourism, and to stimulate tourism in undervalued regions or cities? In other words, further research is needed to take the individual-level psychological mechanisms uncovered in the present chapter to the destination level, so that questions of resident and visitor interests and sustainable development can be optimally addressed.

Our findings exemplify that, with help of data about emotional arousal levels related to digital and physical artifacts in museum exhibits, we could better understand the experience that each visitor, either a tourist or resident, accrues up during a museum visit. Storytelling theory suggests that emotional responses are strongly linked to people's imagination, empathy with characters in a story, their narrative sequences, and memories of the experience. We hope that both our measurement methods and the effectiveness of "The Marquise's Secrets" can be a source of inspiration for DMOs, policymakers, and tourist-recreational facilities.

With accurate, multi-method experience measurements, museums are offered important recommendations to maintain and adapt their museum exhibits to create emotions and memories. Future research could examine how such emotions and memories influence destination image and resulting attitudes or intentions over the long term (Bastiaansen et al., 2018). However, we could imagine that individuals' attitudes or intentions not only relate to a specific museum or location such as the Markiezenhof, but also to the place where the museum is located. Visitors may link their experiences in the Markiezenhof to those with/in the city of Bergen op Zoom, another museum or amenity in this city, a nearby city, or even the wider region. Therefore, information on experience patterns of museum visitors could not only be applied to a specific museum, but also on a larger scale to spread people over cities and regions. While we posit that storytelling has a promising emotional impact in this regard, we also urge destination managers to measure the outcomes of *any* interventions on the outcomes they seek. Accordingly, we hope that both our methods and findings might be helpful to further develop and innovate tourist destinations based on the stories groups of visitors might experience and remember.

References

Baker, C. (2018). Exploring a three-dimensional narrative medium: The theme park as" de sprookjessprokkelaar," the gatherer and teller of stories. Doctoral dissertation, University of Central Florida, Orlando.

Baker, C., Calvi, L., Lub, X., & Hover, M. (2019). Defining a future research agenda on storytelling in tourism, leisure and hospitality: A systematic literature review. *5th World Research Summit for Hospitality and Tourism.*

Bal, M. (2007). Exhibition as film. In S. Macdonald & P. Basu (Eds.), *Exhibition experiments* (pp. 71–93). Blackwell Publishing.

Bal, M., & Van Boheemen, C. (2009). *Narratology: Introduction to the theory of narrative.* University of Toronto Press.

Bargeman, B., Calvi, L., Strijbosch, W., Hover, M., van Waalwijk, J., & Mitas, O. (2022). Interactie met verhalen in een museale context. *Vrijetijdstudies, 40*(2), 13–18.

Bastiaansen, M., Lub, X., Mitas, O., Jung, T. H., Passos Acenção, M., Han, D., Moilanen, T., Smit, B., & Strijbosch, W. (2019). Emotions as core building blocks of an experience. *International Journal of Contemporary Hospitality Management, 31*, 651–668.

Bastiaansen, M., Straatman, S., Driessen, E., Mitas, O., Stekelenburg, J., & Wang, L. (2018). My destination in your brain: A novel neuromarketing approach for evaluating the effectiveness of destination marketing. *Journal of Destination Marketing & Management, 7*, 76–88.

Baumeister, R. F., Vohs, K. D., Nathan DeWall, C., & Zhang, L. (2007). How emotion shapes behavior: Feedback, anticipation, and reflection, rather than direct causation. *Personality and Social Psychology Review, 11*(2), 167–203.

Benedek, M., & Kaernbach, C. (2010). Decomposition of skin conductance data by means of nonnegative deconvolution. *Psychophysiology, 47*(4), 647–658.

Braithwaite, J. J., Watson, D. G., Jones, R., & Rowe, M. (2015). *A guide for analysing electrodermal activity (EDA) & skin conductance responses (SCRs) for psychological experiments.* Technical report.

Bryant, F. B., & Veroff, J. (2017). *Savoring: A new model of positive experience.* Psychology Press.

Calvi, L., Bargeman, B., Hover, M., Waalwijk, J. V., Strijbosch, W., & Mitas, O. (in press). Storytelling as a tool to design museum experiences: The case of the Secret Marquise. In F. Zanella (Ed.), *Multidisciplinary aspects of design: Objects, processes, experiences and narratives.* Springer.

Calvi, L., & Hover, M. (2021). Storytelling for mythmaking in tourist destinations. *Leisure Sciences, 43*(6), 630–643.

Campbell, J., & Moyers, B. (2011). *The power of myth.* Anchor.

Chronis, A. (2012). Tourists as story-builders: Narrative construction at a heritage museum. *Journal of Travel & Tourism Marketing, 29*(5), 444–459.

Eckert, C., Zacher, D., Pechlaner, H., Namberger, P., & Schmude, J. (2019). Strategies and measures directed towards overtourism: A perspective of European DMOs. *International Journal of Tourism Cities, 5*(4), 639–655.

Francis, D. W. (2020). *Excavating Freytag's Pyramid: Narrative, identity and the museum visitor experience* UCL Press.

Fredrickson, B. L. (1998). What good are positive emotions? *Review of General Psychology, 2*(3), 300–319.

Fu, X., Baker, C., Zhang, W., & Zhang, R. (2023). Theme park storytelling: Deconstructing immersion in Chinese theme parks. *Journal of Travel Research, 62*(4), 893–906.

Gabriel, Y. (2000). *Storytelling in organizations: Facts, fictions, and fantasies.* Oxford University Press.

Gilbert, D. (2006). *Stumbling on happiness.* Alfred A. Knopf.

Gordon, R., Ciorciari, J., & van Laer, T. (2018). Using EEG to examine the role of attention, working memory, emotion, and imagination in narrative transportation. *European Journal of Marketing, 52*(1/2), 92–117.

Green, M. C. (2021). Transportation into narrative worlds. In L. B. Frank & P. Falzone (Eds.), *Entertainment-education behind the scenes: Case studies for theory and practice* (pp. 87–101). Palgrave Macmillan/Springer Nature.

Hover, M. (2013). *The Efteling as a "narrator" of fairy tales,* Tilburg University, the Netherlands.

Jelincic, D. (2020). When heritage speaks t-emoticons: Emotional experience design in heritage tourism. In M. Gravari-Barbas (Ed.), *A research agenda for heritage tourism* (pp. 219–234). Edward Elgar.

Kirchberg, V., & Tröndle, M. (2015). The museum experience: Mapping the experience of fine art. *Curator: The Museum Journal, 58*(2), 169–193.

Kono, S., Ito, E., & Gui, J. (2022). Leisure's relationships with hedonic and eudaimonic well-being in daily life: An experience sampling approach. *Leisure Sciences,* 1–20.

Kurtz, J. L., & Lyubomirsky, S. (2013). Happiness promotion: Using mindful photography to increase positive emotion and appreciation. In J. J. Froh & A. C. Parks (Eds.), *Activities for teaching positive psychology: A guide for instructors* (pp. 133–136). American Psychological Association.

Li, S., Sung, B., Lin, Y., & Mitas, O. (2022). Electrodermal activity measure: A methodological review. *Annals of Tourism Research, 96,* 103460.

Macleod, S., Austin, T., Hale, J., & Hing-Kay, O. H. (2018). *The future of museum and gallery design: Purpose, process, perception.* Routledge.

McKee, R. (2010). *Story: Style, structure, substance, and the principles of screenwriting.* HarperCollins.

McKee, R., & Gerace, T. (2018). *Storynomics: Story-driven marketing in the post-advertising world.* Hachette UK.

Mitas, O., Yarnal, C., & Chick, G. (2012). Jokes build community: Mature tourists' positive emotions. *Annals of Tourism Research, 39*(4), 1884–1905.

Mitas, O., Cuenen, R., Bastiaansen, M., Chick, G., & van den Dungen, E. (2020a). The war from both sides: How Dutch and German visitors experience an exhibit of Second World War stories. *International Journal of the Sociology of Leisure, 3*(3), 277–303.

Mitas, O., Mitasova, H., Millar, G., Boode, W., Neveu, V., Hover, M., van den Eijnden, F., & Bastiaansen, M. (2020b). More is not better: The emotional dynamics of an excellent experience. *Journal of Hospitality & Tourism Research, 46*(1), 1096348020957075.

Nabi, R. L., & Green, M. C. (2015). The role of a narrative's emotional flow in promoting persuasive outcomes. *Media Psychology, 18*(2), 137–162.

Petrovic, M., Liedgren, J., & Gaggioli, A. (2022). Beyond influence: Contextualization and optimization for new narrative techniques and story-formats. Perspective paper. *Frontiers in Communication, 7*, 915308.

Reichheld, F. F. (2003). The one number you need to grow. *Harvard Business Review, 81*(12), 46–55.

Richards, G. (2020). Urban studies and the eventful city. In S. J. Page & J. Connell (Eds.), *The Routledge handbook of events* (pp. 273–286). Routledge.

Richards, G., & Duif, L. (2018). *Small cities with big dreams: Creative placemaking and branding strategies*. Routledge.

San Martín, H., & Del Bosque, I. A. R. (2008). Exploring the cognitive–affective nature of destination image and the role of psychological factors in its formation. *Tourism Management, 29*(2), 263–277.

Schwarz, N. (2000). Emotion, cognition, and decision making. *Cognition & Emotion, 14*(4), 433–440.

Skolnick, L. H. (2005). Towards a new museum architecture: Narrative and representation. In S. Macleod (Ed.), *Reshaping museum space* (pp. 118–130). Routledge.

Strijbosch, W., Mitas, O., van Blaricum, T., Vugts, O., Govers, C., Hover, M., Gelissen, J., & Bastiaansen, M. (2021a). Evaluating the Temporal dynamics of a structured experience: Real-time skin conductance and experience reconstruction measures. *Leisure Sciences*, 1–25.

Strijbosch, W., Mitas, O., van Blaricum, T., Vugts, O., Govers, C., Hover, M., Gelissen, J., & Bastiaansen, M. (2021b). When the parts of the sum are greater than the whole: Assessing the peak-and-end-theory for a heterogeneous, multi-episodic tourism experience. *Journal of Destination Marketing & Management, 20*, 100607.

Strijbosch, W., Mitas, O., van Gisbergen, M., Doicaru, M., Gelissen, J., & Bastiaansen, M. (2019). From experience to memory: on the robustness of the peak-and-end-rule for complex, heterogeneous experiences. *Frontiers in Psychology, 10*.

Volgger, M., & Pechlaner, H. (2014). Requirements for destination management organizations in destination governance: Understanding DMO success. *Tourism Management, 41*, 64–75.

Wirtz, D., Kruger, J., Scollon, C. N., & Diener, E. (2003). What to do on spring break? The role of predicted, on-line, and remembered experience in future choice. *Psychological Science, 14*(5), 520–524.

Zajchowski, C. A., Schwab, K. A., & Dustin, D. L. (2016). The experiencing self and the remembering self: Implications for leisure science. *Leisure Sciences*, 1–8.

8 Tourism Interventions

Contributions to Community Resilience

Simone Moretti

Tourism and the VUCA Environment: Between 'Polycrisis' and 'Polysolutions'

A wide range of factors makes it difficult to predict, identify and measure the impacts of any type of intervention in the domain of tourism, as attested by a growing literature on topics such as carrying capacity and overtourism (Butler 2020; Wall 2020), sustainable tourism (Hall 2021; Ruhanen et al. 2019), regenerative tourism (Ateljevic 2020; Bellato et al. 2023), and tourism and climate change (Gössling & Higham 2021; Scott 2021). Among these factors we can certainly mention the multi-stakeholder nature of tourism, the intricate network of power relations behind its governance, and the influence of external and unpredictable events and disruptions. Therefore, tourism scholars introduced constructs such as 'chaos' and 'complexity' as alternative frameworks to look at the tourism phenomena (Altinay & Arici 2022; Hartman 2021; Heylighen 2002), due to deficiencies in understanding the tourism systems by only using frameworks based on linear changes or stability assumptions (Faulkner & Russell 1997).

For instance, Hartman (2016) proposed the application of the conceptual framework of Complex Adaptive Systems (CAS) to study tourism phenomena. The author describes how tourism areas can be seen as complex systems of interrelated products, sectors and institutions, carrying out actions that are very closely tied to other elements, agents and their actions, generating circular cause-effect consequences, resulting in a complex and volatile environment that is constantly changing (Farrell & Twining-Ward 2004; Hartman 2016; McDonald 2009). Therefore, tourism systems are constantly involved in a continuous process of adaptation, responding to and anticipating changes that might challenge or put these systems at risk (e.g., rise of sharing economy, climate change, overtourism, COVID-19 pandemic, etc.).

Further elaborating on these dynamic processes, Calvi and Moretti (2020) observed how tourism systems and their actors are now operating in what has been named as a 'Volatile, Uncertain, Complex, and Ambiguous' environment, in short, a 'VUCA environment'. The acronym VUCA has its origin in the military vocabulary, where it is used to describe an uncertain environment with threats at every step, a situation in which a conflict is almost impossible to predict (Minciu et al. 2020).

DOI: 10.4324/9781003449027-10

Such situations are 'volatile' as they can rapidly change, 'uncertain' as it is difficult to predict how they will evolve due to the influence of many factors, 'complex' as they involve many interconnected aspects and actors, and 'ambiguous' as they can be interpreted from different perspectives and often actors have to make decisions based on incomplete or partial information. The same concept also describes the current societies, in which several socio-economic phenomena take place, including tourism. In order to survive and prosper in the current unstable environment, socio-economic actors and communities must learn how to deal with changes and unexpected situations typical of a VUCA environment. This justifies the increasing importance attributed to the concept of resilience, generally defined by Jones and Comfort (2020: 2) as 'the ability to withstand or to bounce back from adversity and disruptions', while Adger (2000: 347), defines social resilience as 'the ability of groups or communities to cope with external stresses and disturbances as a result of social, political and environmental change'. Resilience is sometimes discussed in relation to tourism systems, but often, and most importantly, it is described in relation to the resilience of communities and society as a whole (Kirmayer et al. 2009; Koliou et al. 2018).

In the current academic debate, tourism is often seen as intertwined with 'polycrises' and specifically identified as a contributor to a global crisis caused by many different problems happening at the same time, producing together big negative impacts (e.g., climate change, biodiversity loss, pandemics, social injustices, etc.). A non-ideological approach, while recognizing critical aspects, should not neglect the potential of tourism in strengthening social justice and playing a fair, although small, role in the mitigation of global crises. In this perspective, tourism can contribute to what we might define as 'polysolutions': networks of singularly marginal contributions, able to jointly determine a bigger positive response to polycrises.

To support this argument, this chapter makes use of Bourdieu's theory of capital (Bourdieu 1986), highlighting the strong connection existing between the accumulation of different forms of capital and community resilience. The empirical case study of 'Migrantour' is then used to demonstrate how tourism interventions can contribute to the accumulation of different forms of capital and, therefore, strengthen the resilience of tourism destinations and their sub-communities involved in tourism. Relevant insights concerning Migrantour have been retrieved from a case study conducted for the EU-funded Horizon 2020 research project 'SmartCulTour'.

Bourdieu's Theory of Capital

Pierre Bourdieu, a French sociologist and anthropologist of the twentieth century, challenged the notion of capital interpreted as a one-dimensional economic entity. Instead, he proposed a multifaceted approach that encompassed various forms of capital, each contributing to shaping social realities (Bourdieu 1986). Bourdieu's theory of capital recognizes the impossibility of understanding 'the social world without acknowledging the role of "capital" in all its forms, not just the one form recognized by economic theory' (Bourdieu 1986: 22).

While from the perspective of economic theory, the term capital originally refers to 'an accumulated sum of money, which could be invested in the hope of future profits' (Field 2003: 12), Bourdieu (1986) argues that economic capital, despite its visual and material tangibility, is not the only form of capital that is relevant and significant, even in a capitalism-oriented society. Other forms of capital, such as social capital (Bourdieu 1986; Coleman 1990; Field 2003; Putnam 2000), cultural capital (Bourdieu 1984) and symbolic capital (Bourdieu 1986) must be considered to outline a more comprehensive picture of how societies work. Each form of capital can be accumulated over time and has the potential to reproduce itself in identical or expanded forms (Bourdieu 1986).

Economic capital refers to a wide range of material assets that are directly convertible into money and may be institutionalized in the form of property rights (Bourdieu 1986). Therefore, economic capital includes all kinds of accumulated wealth as material resources. For example, financial resources, goods, land or property ownerships, etc. (Pinxten & Lievens 2014).

Bourdieu sees cultural capital as the combination of knowledge, expertise and experience embodied in individuals, accumulated during their life, but also derived from one's social origins (Bourdieu 1986, 1991). According to Bourdieu's view, cultural capital can exist in three different forms. Firstly, 'embodied cultural capital' values all the long-lasting dispositions, such as ethnicity, religion, or family background, acquired from early childhood, learned from family, or gained through personal experiences and socializations (Bourdieu 1977, 1984, 1986). Secondly, 'objectified cultural capital' attributes value and power to cultural artifacts and goods, such as pictures, books, dictionaries, instruments, etc., including the individual exposure to these goods (for example, through cultural experiences by visiting museums, art galleries, theatres, etc.) (Bourdieu 1977, 1984, 1986). Finally, Bourdieu (1986, 1984, 1977) defines 'institutionalized cultural capital' as cultural competency earned through formal degrees, certifications, and qualifications obtained through the educational system.

In a Bourdieusian perspective, social capital is 'the aggregate of the actual or potential resources which are linked to possession of durable network of more or less institutionalized relationships of mutual acquaintance and recognition' (Bourdieu, 1986: 244). According to Field (2003), those who possess relatively high levels of economic and cultural capital also tend to have high levels of social capital, although subordinate groups with strong social capital as a whole can thrive despite the absence of economic and cultural capital.

Symbolic capital refers to the possession of symbolic resources, thus consisting of the 'prestige and renown attached to a family and a name' (Bourdieu 1977: 179). It can be argued that symbolic capital only exists in the eyes of others.

Through the years, Bourdieu's contribution found a wide range of applications in tourism studies. For instance, scholars adopted Bourdieu's forms of capital as a framework to study tourism entrepreneurial practices (Çakmak et al. 2018); to understand the role of groups of stakeholders in tourism, such as tour guides (Garner 2016); or to study specific forms of tourism, such as community-based tourism (Pramanik et al. 2019; Wulandari et al. 2022).

Community Resilience

The complexity of societal systems in which tourism happens and the perspective of operating in a VUCA environment characterized by sudden and unexpected changes, disruptions and crises, justify the increased importance assumed by the concept of resilience. The concept itself derives from the physics of materials and has been adapted by multiple disciplines, including engineering, environmental sciences, economics, psychology and sociology (Kirmayer et al. 2009; Koliou et al. 2018). Its widespread diffusion testifies the recognition that not all threats, disruptions and crises can actually be avoided or anticipated. Hence, the need to minimize disturbances and make sure that adaptive and transformative approaches are embraced, allowing societies, organizations, communities and individuals to engage in a long-term evolution path (Elmqvist 2014; Matyas & Pelling 2015; Renschler et al. 2010; Sharifi 2016) or, in other words, to ensure their resilience. Due to the conceptual application in a variety of domains and disciplines, it is not surprising that resilience has been defined from multiple perspectives and that there is no consensus on a single conceptual definition. Nevertheless, for the objective of this chapter, it seems particularly relevant to consider the definition of Adger (2000: 347), who defines social resilience as 'the ability of groups or communities to cope with external stresses and disturbances as a result of social, political and environmental change'.

This chapter, using the lenses of Bourdieu's theory of capital, aims to understand how certain tourism interventions can strengthen the resilience of communities that are hosting visitors. Nevertheless, understanding community resilience also entails understanding the resilience of the individuals that are part of a certain community, as 'understanding the characteristics of resilient individuals can help identify those features of communities that enable or facilitate individuals to thrive' (Kirmayer et al. 2009: 63). In other words, if several individuals belonging to a community develop individual resilience, this might contribute to making the entire community more resilient, since individuals can more effectively work together when responding to stresses, disruptions and crises.

Similarly to the broader concept of resilience, a uniform consensus has not been reached on what community resilience is and how it should be defined, leading to mixed definitions appearing in the scientific literature, policies and practice. Castleden et al. (2011: 370) described community resilience as 'a capability (or process) of a community adapting and functioning in the face of disturbance'. Others have defined community resilience more as a network of features of a community, such as 'household relationships, levels of education and literacy, employment-seeking behaviours, social support networks, ability to seek support services, sense of communal safety and hope, and physical security measures' (Ahmed et al. 2004: 393). Considering the scope of this study, the definition provided by Magis (2010: 402) is of particular interest, where community resilience is seen as

the existence, development and engagement of community resources by community members to thrive in an environment characterised by change, uncertainty, unpredictability, and surprise. Members of resilient communities intentionally develop personal and collective capacity that they engage to respond to and influence change, to sustain and renew the community, and to develop new trajectories for the communities' future.

Elements of Community Resilience

The literature on community resilience offers contributions aimed at identifying elements, factors, criteria and conditions that, when existing within a community, contribute to building up its resilience. Jordan and Javernick-Will (2012) conducted an extended review of definitions of community resilience to assess the most commonly used measurement of this concept. Indicators such as the level of education, place attachment and access to information emerged from their analysis. According to Kirmayer et al. (2009), the ability of a system to return to a steady state or adapt by transforming the system itself, requires a variety of competencies in the fields of communication, emotion, spirituality and community relationships. Ahmed et al. (2004: 393) identified the following as key dimensions of community resilience: 'household relationships, levels of education and literacy, employment-seeking behaviours, social support networks, ability to seek support services, sense of communal safety and hope, and physical security measures'.

Sharifi (2016) analysed existing 'community resilience assessment tools' and identified five main dimensions of community resilience: environmental, social, economic, built environment- infrastructure and institutional. Each dimension is divided into sub-dimensions and further divided into several resilience criteria. Some of these dimensions and criteria are particularly interesting and are reported in Table 8.1.

Similarly, Patel et al. (2017) carried out a comprehensive literature review on community resilience and identified several factors that are crucial to building up community resilience:

- Local knowledge: knowledge (current available information and previous experiences from the past) and understanding of strengths and vulnerabilities increases the resilience of a community as, in case of disruptions, individuals are already informed about significant elements the community can leverage to implement adaptive mechanisms. Training and education represent important tools to improve local knowledge;
- Community network and relationships: community resilience is higher when its members are well connected, forming a proper social network and a cohesive whole. Trust and shared values can strengthen a community network and build up community resilience, as in case of disruptions it will be easier to share knowledge and create synergies to implement adaptive mechanisms;
- Communication: communication networks are critical to strengthen local knowledge and develop community networks. Effective communication occurs when common meanings are shared within a community, creating a common understanding, and if the community is provided with opportunities for open

Table 8.1 Dimensions, sub-dimensions and criteria to assess community resilience

Dimensionsons of Community Resilience	Sub-Dimensions	Examples of criteria to assess resilience
Social	Equity & Diversity	Fair access to basic needs and services, diverse workforce, gender and ethnic equality
	Local Culture	Learning from the past, cultural and historical preservation, cultural knowledge
	Community bond & social Support	Civic engagement in social networks connectedness across community groups, empowerment of vulnerable groups, place attachment, sense of community, trust
	Safety & Well being	Physical and Psychological health, crime prevention
	Social Structure	Diversity of skills and expertise
Economic	Structure	Employment rate, Income, employable skills and qualifications
	Security	Individual and community savings, Social welfare
	Dynamism	Entrepreneurialism, public-private partnerships, locally owned businesses, diverse economic structure
Institutional	Education & Training	Education and training system, communication skills, capacity building
	Collaboration	Cross-sector or cross-groups collaborations and partnerships, knowledge and information transfer
	Leadership & Participation	Multi-stakeholder planning and decision-making, transparency, long-term vision,strong leadership

Source: Author's adaptation based on Sharifi (2016)

dialogue. When facing disruptions, crises or unexpected changes, effective communication networks and common understandings are crucial to promptly share information;

- Governance and leadership: they shape how communities handle crises and unexpected situations, implementing adaptive mechanisms. Participation and representation of the local community in strategic planning of adaptive mechanisms are also important to strengthen community resilience;
- Resources: a higher level of resources is generally supposed to determine higher levels of resilience, as well as a fair and equitable distribution of available resources within the community and its groups, including more tangible elements (e.g., water, energy, food etc.) but also less tangible aspects such as human, financial and social resources;
- Economic investments: disruption and crisis might cause unavoidable short-term economic costs that, if not appropriately addressed, can affect the community in the long term. Moreover, certain elements can strengthen the resilience of the community, such as fair distribution and access to financial resources, cost-effective investments in infrastructures and facilities, diversity of

economic resources, and proactive investments to revitalize the job market and stimulate economic growth;

* Mental outlook: this refers to the attitudes, feelings and views a community and its individuals develop when facing uncertainties that are typical of a disruption. The mental outlook of a community is crucial in shaping the willingness and ability of community members to carry on and implement adaptive mechanisms.

Bridging Capitals and Community Resilience

By comparing Bourdieu's forms of capital and the mentioned indicators of community resilience, the existence of a strong connection between them becomes evident. These links will be now further explored from a conceptual perspective and then investigated through the empirical case study of Migrantour, providing an example of how tourism interventions can support the accumulation of different forms of capital, therefore strengthening community resilience in tourism destinations.

Bourdieu defined social capital as 'the aggregate of the actual or potential resources which are linked to possession of durable network of more or less institutionalized relationships of mutual acquaintance and recognition' (Bourdieu, 1986: 244). The possession of a durable network and established relationships seems very much in line with some of the social and institutional resilience criteria mentioned by Sharifi (2016) specifically concerning community bonds and social support (e.g., civic engagement in social networks, connectedness across community groups, empowerment of vulnerable groups, sense of community, trust), collaboration (e.g., cross-sector or cross-group collaborations and partnerships, knowledge and information transfer), and participation (e.g., multi-stakeholder planning and decision-making). Social capital also seems well connected with the 'community network' and 'relationships' elements mentioned by Patel et al. (2017), supporting the idea that community resilience is higher when its members are forming a proper social network. Similarly, a et al. (2009) and Ahmed et al. (2004) recognize community relationships and social support networks as crucial elements for community resilience. The conceptual links between social capital and community resilience indicators are summarized in Table 8.2.

Following a similar approach, cultural capital defined as a combination of knowledge, expertise and experience embodied in individuals, accumulated during their life but also derived from one's social origins (Bourdieu 1986, 1991), appears strongly associated with some of the social, economic and institutional criteria mentioned by Sharifi (2016). Particularly, we can draw clear connections with sub-dimensions like local culture (e.g., learning from the past, cultural and historical preservation, cultural knowledge), social support (e.g., place attachment, sense of community), social structure (e.g., diversity of skills and expertise), education and training (e.g., education and training system, communication skills, capacity building), and economic structure (e.g., employable skills and qualifications). Cultural capital also seems closely connected with elements mentioned by Patel et al. (2017), such as 'local knowledge', 'communication' and, to some extent, 'mental outlook'. Similarly, Jordan and Javernick-Will (2012) suggest indicators, such

Table 8.2 Conceptual links between social capital and community resilience

Conceptual links based on Shariffs(2016) dimensions and sub-dimensions	Conceptual links based on Patel et al.(2017)	Conceptual links based on other contributions
Social dimensions: • Community bond and Social Support (e.g., Civic engagement in social networks, Connectedness across community groups, empowerment of vulnerable groups, Sense of community, trust) • Collaboration (e.g., Cross-sector or cross-groups collaborations and partnerships, Knowledge and information transfer) **Institutional dimensions:** • Participation (e.g., multi-stakeholder Planning and decision-making)	• Community network • Relationships	• Community relationship • Social support networks

as the level of education, place attachment and access to information, as crucial elements for community resilience. Higher levels of education and literacy were also indicated by Ahmed et al. (2004) as important contributors to community resilience. The conceptual links between cultural capital and community resilience indicators are summarized in Table 8.3.

Bourdieu refers to economic capital as a wide range of material assets including all kinds of accumulated wealth as material resources. Here the close connection with the economic dimension mentioned by Sharifi (2016) is noticeable for all of its sub-dimensions, namely the economic structure (e.g., employment rate, income), security (e.g., individual and community savings) and dynamism (e.g., entrepreneurialism, locally owned businesses, diverse economic structure). The connection with the contribution of Patel et al. (2017) is mostly concerning the availability of resources (especially tangible ones) and economic investments (e.g., fair access and distribution of financial resources; cost-effective investments in infrastructures and facilities; diversity of economic resources; proactive investments to revitalize the job market, distribute economic aid or stimulate economic growth).

Symbolic capital refers to the possession of symbolic resources such as 'prestige and renown attached to a family and a name' (Bourdieu 1977: 179). Its connection with resilience enablers may be less intuitive. Nevertheless, it could be expected that symbolic capital could facilitate community bonds and social support (e.g., individuals with recognized prestige across the community might be invested with the role of connectors between community groups). Symbolic capital might also be an element of economic dynamism (e.g., entrepreneurialism, public-private partnerships, locally owned businesses) or contribute to leadership and participation (e.g., strong leadership). Symbolic capital can also be controversial and actually hinder elements of community resilience if misused, for example, to hold and retain power (e.g., prestige can turn into elitism and hinder community bonds, social support and community participation). Symbolic capital might also be seen

Table 8.3 Conceptual links between cultural capital and community resilience

Conceptual links based on Sharifi's (2016) dimensions and sub-dimensions	*Conceptual links based on Patel et al. (2017)*	*Conceptual links based on other contributions*
Social dimension: • Local culture (e.g., Learning from the Past, Cultural and historical Preservation, cultural Knowledge) • Social support (e.g., Place attachment, Sense of community) • Social Structure (e.g., diversity of skills and expertise) **Economic dimension:** • Economic structure (e.g., employable skills and qualifications) **Institutional dimension:** • Education and Training (e.g., education and training system, Communication skills, capacity building)	• Local knowledge • Communication • Mental outlook	• Level of education and literacy • Place attachment • Access to information

Table 8.4 Conceptual links between economic capital and community resilience

Conceptual links based on Sharifi's (2016) dimensions and sub-dimensions	*Conceptual links based on Patel et al. (2017)*	*Conceptual links based on other contributions*
Economic dimension: Structure (e.g., employment rale, income). Security (e.g.. Individual and community savings) Dynamism (e.g., entrepreneurialism, locally owned business, diverse economic structure)	• Avail lability of resources • Avail lability of investments	

in connection with the 'mental outlook' mentioned by Patel et al. (2017). For instance, individuals or community groups possessing symbolic prestige and recognition might effectively assume a leading role in shaping the willingness and ability of community members to implement adaptive mechanisms in front of disruptions and uncertainties.

Concluding, the analysis supports the existence of a strong conceptual connection between different forms of capital and community resilience, suggesting that individual and community accumulation of different forms of capital play a crucial role in building up and strengthening community resilience. The following section supports this argument by identifying and unravelling these connections by looking at the case study of a specific tourism intervention (Migrantour).

Table 8.5 Conceptual links between symbolic capital and community resilience

Conceptual links based on Sharif's (2016) dimensions and sub-dimensions	Conceptual links based on Patel et al. (2017)	Conceptual links based on other contributions
Social dimension: • Community bond and social support (e.g., Individuals with recognized prestige across the community might be invested with the role of connectors between community groups) **Economic dimension:** • Dynamism (e.g., entrepreneurialism, public-private partnerships, locally owned businesses) **Institutional dimension:** • Leadership and participation (e.g., strong leadership)	• Mental outlook	

Tourism Interventions, Forms of Capital and Community Resilience: The Case of Migrantour

The considered tourism intervention consists of a new experience (intercultural walks), named 'Migrantour', offered to tourists, but also to residents and school students. Migrantour originates from the encounter between an anthropologist interested in the connection between migration and tourism and a tour operator in responsible tourism (Viaggi Solidali). The 'experimental' phase of the intervention took place in Turin (Italy) and was supported by two NGOs (Oxfam Italia and ACRA-CCS). Since then, Migrantour expanded first in Italy, then generated a network involving several cities in Europe through the involvement of local associations and tour operators.

The initiators of the intervention observed how several European cities are experiencing challenges in the socio-cultural integration of migrants and how these challenges translate into socio-demographic developments leading to neighbourhoods mostly inhabited by specific groups of migrants. These neighbourhoods are often seen by tourists (and other residents) as dangerous and unattractive. Even the socio-cultural heritage of these 'new locals' is often looked at with stigma or even neglected. Within these contexts, Migrantour offers 'intercultural walks' facilitated by an 'intercultural companion' (a local resident with a migration background), exploring neighbourhoods that have been shaped and influenced by significant migration flows. These walks involve interactions with other locals with migrant backgrounds (e.g., owners of shops, members of the religious community, etc.).

The case study conducted on Migrantour revealed substantial contributions of the intervention in terms of social, cultural, economic and symbolic capital accumulation for disadvantaged sub-groups of the involved communities, such as locals with a migration background involved as 'intercultural companions'.

Migrantour, Social Capital and Community Resilience

As a result of their involvement in Migrantour, intercultural companions reported a higher level of integration in the local community, the feeling of being more active as citizens, and the opportunity to establish and develop a network of relationships based on mutual recognition. As intercultural walks are not only experienced by tourists but also by other residents, the walks promote bonding between different groups of a larger urban community and even allow local small businesses run by migrants to get in touch with other components of a larger community. On a small scale, Migrantours have been recognized as promoting relational dynamics between residents with a migrant background, native residents and visitors. Therefore, the intervention clearly plays a role in terms of the accumulation of social capital as defined by Bourdieu.

The conceptual links previously established (see Table 8.2) explain how the accumulation of social capital determined by Migrantour contributes to community resilience: not only does it strengthen the community network and the relationships between its components, but it also positively influences community bonding (e.g., civic engagement in social networks, connectedness across community groups, empowerment of vulnerable groups) and the level of collaboration within the community (e.g., cross-sector or cross-group collaborations and partnerships, knowledge and information transfer). As mentioned by Patel et al. (2017), community resilience is higher when its members are well connected and form a proper social network, because in cases of disruption it will be easier to create synergies necessary to efficiently implement adaptive mechanisms. Looking at resilience at the individual level, Migrantour's intercultural companions will benefit from their increased connectedness with the community networks in case of future disruptions and crisis. For instance, in the event of significant socio-demographic changes in the neighbourhood, they will have developed useful connections within and outside their sub-group, mitigating possible social alienation. In the event of changes requiring them to seek a new job, they could rely on a network that might facilitate access to new opportunities. In general, networks and connections are crucial to access relevant information and opportunities during a wide range of disruptions, changes and disasters.

Migrantour, Cultural Capital and Community Resilience

Migrantour also increases the cultural capital of intercultural companions, in the form of knowledge, expertise and experience embodied in individuals. They undergo a training programme, equipping them with a variety of useful soft skills and cultural-historical knowledge and expertise, which they also develop throughout their entire experience with Migrantour. They reported increased self-esteem and self-realization of their own capabilities and an increased sense of belonging.

Therefore, the intervention plays a role in terms of the accumulation of cultural capital in all the three forms mentioned by Bourdieu: embodied, objectified and institutionalized.

The conceptual links previously established (see Table 8.3) explain how the accumulation of cultural capital determined by Migrantour contributes to several elements of community resilience, such as reinforcement of local knowledge and culture (e.g., learning from the past, cultural and historical preservation, cultural knowledge) as well as place attachment and sense of community. Within the community, the diversity of skills and expertise is increased through education and training, with the result being increased employable skills and qualifications for vulnerable components of the local community. As mentioned by Patel et al. (2017), training and education represent important tools to improve local knowledge and increase individual and community preparedness to face disruptions and crises. Considering the impact on individual resilience, Migrantour's intercultural companions can benefit from the developed skills and expertise to access and perform other jobs in the future, should this change be determined by unexpected disruptions. It could be objected that any type of job experience would produce similar benefits. Nevertheless, it can be argued that tourism-related job experiences, even at entry-level, often allow the development of interpersonal, communication, language and soft skills, which might represent a valuable wealth of experiences if compared to entry-level jobs in alternative fields (e.g., agriculture or factory work). These types of expertise and professionalization are also generally more easily transferable on the job market, which is very important in the context of a VUCA environment.

Migrantour, Economic Capital and Community Resilience

To some extent, Migrantour also promotes the accumulation of economic capital. Part-time jobs are created, mostly for the intercultural companions, providing an income that supports often marginalized individuals, while developing their skills and professionalization. The tour operators commercializing the Migrantours on a local level also accumulate economic capital, as well as (to some extent) small businesses that are involved in the intercultural walks (also often owned by residents with a migrant background).

The conceptual links previously established (see Table 8.4) explain how the accumulation of economic capital determined by Migrantour contributes to elements of community resilience, as it positively influences employment rates and income, security, and economic dynamism. As suggested by Patel et al. (2017), higher availability of resources (including financial means) generally leads to higher levels of resilience. Ensuring a fair allocation of resources within the community also contributes to community resilience. Disruption and crisis might cause unavoidable short-term direct and indirect economic costs; availability of fairly distributed economic capital can help avoid long-term impact on the entire community. For example, at an individual level, the income of intercultural companions can mitigate the short-term economic impact of a crisis determining the unemployment of another family member, or the sudden increase of electricity costs, or changes in the social welfare system. The more individuals are in a position to absorb these economic disruptions, the more the entire community is resilient.

Migrantour, Symbolic Capital and Community Resilience

The accumulation of symbolic capital in terms of 'prestige and renown attached to a family or a name' is probably the least visible form of capital generated through an intervention such as Migrantour. Nevertheless, a broader interpretation of this concept allows the identification of a symbolic capital accumulation in terms of visibility and recognition that Migrantour brought to neighbourhoods in which intercultural walks are organized. This sometimes stimulated the presence of other projects and associations aiming at socio-economic improvements and attracted the attention of municipalities, leading to urban regeneration initiatives and improved facilities and infrastructures.

The conceptual links previously established (see Table 8.5) shows how the accumulation of symbolic capital determined by Migrantour contributes to community resilience, as the mentioned symbolic capital facilitates a reinforcement of the social support system available for the community members and, indirectly, it strengthens the participation and the leadership towards a future vision for the community, through the attraction of different types of actors interested in future developments. The accumulation of symbolic capital makes the community more prepared and supported when facing unexpected changes and disruptions, as more actors will have a stake in it.

Conclusion

The current academic debate often sees tourism as a contributor to 'polycrises'. While recognizing critical aspects, a non-ideological approach should not neglect the potential of tourism in strengthening social justice and playing a fair role in the mitigation of global crises, contributing to what could be defined as 'polysolutions'. The ability of society and communities to deal with and respond to polycrises by implementing polysolutions also depends on their level of resilience in terms of ability to withstand, cope or bounce back from adversity and disruptions resulting from social, political and environmental changes. As previously mentioned, in order to survive and prosper in the current VUCA environment, socio-economic actors and communities must learn how to deal with changes and unexpected situations where their survival depends on their level of resilience.

This chapter argues that well-designed tourism interventions facilitate the accumulation of different forms of capital within communities that are hosting visitors and those capitals serve as contributors in building up the resilience of tourism destinations and their communities. To support this argument, conceptual links between elements of community resilience and Bourdieu's forms of capital were drawn. Then, the case of Migrantour was considered, providing an empirical example of how a tourism intervention can generate different forms of capital and how these capitals influence elements of community resilience.

The embraced approach provides valuable insights for tourism practitioners and governance actors. The direct impacts in terms of social, cultural and symbolic capital and the indirect effect on community resilience should always be considered when designing a certain tourism intervention or when anticipating the potential results. Evaluating the impacts on different forms of capitals and on community resilience gives a more comprehensive vision on tourism impacts, providing tourism operators with the opportunity to better communicate the societal value of their activities, beyond just the economic dimension. Moreover, governance frameworks should encourage tourism operators to maximize their impacts on social, cultural and symbolic capital, which would positively affect communities' resilience and strengthen their capacity to withstand and cope with disruptions resulting from social, political and environmental changes.

Therefore, this chapter also provides a response to the increasing number of voices calling for redefining the concept of success in tourism. This means that tourism impacts should be evaluated beyond the hard numbers of tourist arrivals and tourist expenditure, but also by considering the impacts on other societal values, such as the contribution in terms of resilience of destinations and their communities.

References

Adger, W. N. (2000). Social and ecological resilience: Are they related? *Progress in Human Geography, 24*(3), 347–364.

Ahmed, R., Seedat, M., Van Niekerk, A., & Bulbulia, S. (2004). Discerning community resilience in disadvantaged communities in the context of violence and injury prevention. *South African Journal of Psychology, 34*(3), 386–408.

Aldrich, D. P. (2012). *Building resilience: Social capital in post-disaster recovery.* University of Chicago Press.

Altinay, L., & Arici, H. E. (2022). Transformation of the hospitality services marketing structure: A chaos theory perspective. *Journal of Services Marketing, 36*(5), 658–673.

Ateljevic, I. (2020). Transforming the (tourism) world for good and (re)generating the potential 'new normal'. *Tourism Geographies, 22*(3), 467–475.

Bellato, L., Frantzeskaki, N. & Nygaard, C. A. (2023). Regenerative tourism: A conceptual framework leveraging theory and practice. *Tourism Geographies, 25*(4), 1026–1046.

Bourdieu, P. (1977). *Outline of theory of practice.* Cambridge University Press.

Bourdieu, P. (1984). *Distinction: A social critique of the judgement of taste.* Routledge and Kegan Paul.

Bourdieu, P. (1986). The forms of capital. In J. Richardson (Ed.), *Handbook of theory and research for the sociology of education* (pp. 241–258). Bloomsbury.

Bourdieu, P. (1991). *Language and symbolic power* (J. B. Thompson, Ed.). Polity Press.

Butler, R. W. (2020). Tourism carrying capacity research: A perspective article. *Tourism Review, 75*(1), 207–211.

Calvi, L., & Moretti, S. (2020). *Future of cultural tourism for urban and regional destinations.* Deliverable D2.2 of the Horizon 2020 project SmartCulTour (GA number 870708), October. http://www.smartcultour.eu/deliverables/.

Çakmak, E., Lie, R. & McCabe, S. (2018). Reframing informal tourism entrepreneurial practices: Capital and field relations structuring the informal tourism economy of Chiang Mai. *Annals of Tourism Research, 72*, 37–47.

Castleden, M., McKee, M., Murray, V. & Leonardi, G. (2011). Resilience thinking in health protection. *Journal of Public Health, 33*(3), 369–377.

Coleman, J. S. (1990). *Foundations of social theory.* Harvard University Press.

Elmqvist, T. (2014). Urban resilience thinking. *Solutions, 5*(5), 26–30.

Farrell, B. H., & Twining-Ward, L. (2004). Reconceptualizing Tourism. *Annals of Tourism Research, 31*(2), 274–295.

Faulkner, B., & Russell, R. (1997). Chaos and complexity in tourism: In search of a new perspective. *Pacific Tourism Review, 1*(2), 93–102.

Field, J. (2003). *Social capital.* Routledge.

Garner, R. (2016). Insecure positions, heteronomous autonomy and tourism-cultural capital: A Bourdieusian reading of tour guides on BBC Worldwide's *Doctor Who* Experience Walking Tour. *Tourist Studies, 17*(4), 426–442.

Gössling, S., & Higham, J. (2021). The low-carbon imperative: Destination management under urgent climate change. *Journal of Travel Research, 60*(6), 1167–1179.

Hall, C. M. (2021). Constructing sustainable tourism development: The 2030 agenda and the managerial ecology of sustainable tourism. In K. A. Boluk, C. T. Cavaliere & F. Higgins-Desbiolles (Eds.), *Activating critical thinking to advance the sustainable development goals in tourism systems* (pp. 198–214). Routledge.

Hartman, S. (2016). Towards adaptive tourism areas? A complexity perspective to examine the conditions for adaptive capacity. *Journal of Sustainable Tourism, 24*(2), 299–314.

Hartman, S. (2021). Adaptive tourism areas in times of change. *Annals of Tourism Research, 87*, 102987.

Heylighen, F. (2002). The science of self-organization and adaptivity. *The Encyclopedia of Life Support Systems.* EOLSS Publishers.

Jones, P., & Comfort, D. (2020). The role of resilience in research and planning in the tourism industry. *Athens Journal of Tourism, 7*(1), 1–16.

Jordan, E., & Javernick-Will, A. (2012). Measuring community resilience and recovery: A content analysis of indicators. *Construction Research Congress 2012: Construction challenges in a flat world*, 2190–2199.

Kirmayer, L. J., Sehdev, M., Whitley, R., Dandeneau, S. F. & Isaac, C. (2009). Community resilience: Models, metaphors and measures. *International Journal of Indigenous Health, 5*(1), 62–117.

Koliou, M., van de Lindt, J. W., McAllister, T. P., Ellingwood, B. R., Dillard, M. & Cutler, H. (2018). State of the research in community resilience: progress and challenges. *Sustainable and Resilient Infrastructure, 5*(3), 131–151.

Magis, K. (2010). Community resilience: An indicator of social sustainability. *Society and Natural Resources, 23*(5), 401–416.

Matyas, D., & Pelling, M. (2015). Positioning resilience for 2015: The role of resistance, incremental adjustment and transformation in disaster risk management policy. *Disasters, 39*(s1), s1–s18.

McDonald, J. R. (2009). Complexity science: An alternative world view for understanding sustainable tourism development. *Journal of Sustainable Tourism, 17*(4), 455–471.

Minciu, M., Berar, F. A. & Dobrea, R. C. (2020). New decision systems in the VUCA world. *Management and Marketing, 15*(2), 236–254.

Patel, S. S., Rogers, M. B., Amlôt, R. & Rubin, G. J. (2017). What do we mean by 'community resilience'? A systematic literature review of how it is defined in the literature. *PLoS Currents, 9*.

Pinxten, W., & Lievens, J. (2014). The importance of economic, social and cultural capital in understanding health inequalities: Using a Bourdieu-based approach in research on physical and mental health perceptions. *Sociology of Health & Illness, 36*(7), 1095–1110.

Pramanik, P. D., Ingkadijaya, R. & Achmadi, M. (2019). The role of social capital in community based tourism. *Journal of Indonesian Tourism and Development Studies*, *7*(2), 62–73.

Putnam, R. D. (2000). *Bowling alone: The collapse and revival of American community.* Simon & Schuster.

Renschler, C., Frazier, A., Arendt, L., Cimellaro, G. P., Reinhorn, A. & Bruneau, M. (2010). Developing the 'PEOPLES' resilience framework for defining and measuring disaster resilience at the community scale. In *Proceedings of the 9th US National and 10th Canadian Conference on Earthquake Engineering (9USN/10CCEE)*, 25–29.

Ruhanen, L., Moyle, C. L. & Moyle, B. (2019). New directions in sustainable tourism research. *Tourism Review*, *74*(2), 245–256.

Scott, D. (2021). Sustainable tourism and the grand challenge of climate change. *Sustainability*, *13*(4), 1966.

Sharifi, A. (2016). A critical review of selected tools for assessing community resilience. *Ecological Indicators*, *69*, 629–647.

Wall, G. (2020). From carrying capacity to overtourism: A perspective article. *Tourism Review*, *75*(1), 212–215.

Wulandari, W. O. S. A., Dirman, L. O. & Aso, L. (2022). Social practices in tourism village development at Loghiya Village, Indonesia. *International Journal of Social Science and Human Research*, *5*(8).

9 The Urban Leisure and Tourism Labs as Incubators for Sustainable Tourism Interventions

Sharing Insights from Educational Research Perspectives on Regenerative Placemaking in Amsterdam and Rotterdam

Roos Gerritsma and Donagh Horgan

Background: The need for a more holistic (collective) approach to sustainable tourism development

Despite the reframing of destination management discourse during the recent global shutdown, moving towards more sustainable tourism remains a challenge for many destinations. In urban areas, many intractable problems exist that necessitate networked collaboration across the entire destination ecosystem (Koens & Milano 2023). Wicked problems are well documented and are often interlinked with other complex urban challenges facing destinations and their decision-makers (Urry 2016; Huijbens & Jóhannesson 2020). These range from disruptions to quality of life and place for local populations and depletion of resource and cultural heritage, to the ecological impacts on destinations, and the planet as a whole. This has given rise to an emerging degrowth discourse in tourism that remains incompatible with the ambitions of economic development actors in destinations (Higgins-Desbiolles & Everingham 2022). In the Netherlands, and specifically in Amsterdam, overtourism had been widely criticised prior to the pandemic, with policymakers again introducing ways to limit overcrowding and antisocial behaviour (Gerritsma 2019; Koens et al. 2018; Oskam & Wiegerink 2020).

Taking the case of Amsterdam in particular – which like Venice and Lisbon suffers the impacts of overtourism in areas such as housing and mobility – communities are experiencing participation fatigue, frustrated by inaction and the lack of concrete outcomes from engagement programmes such as 'Stad in Balans' (Horgan & Gerritsma 2023). Engaged research into overtourism and social exclusion for the SMARTDEST project revealed, among other insights, that there does not seem to be a clear overview of all those that are involved in (co)designing new tourism policies, nor visibility as to how decisions are taken, as well as lack of a long-term vision for how to facilitate more open participation trajectories (Gerritsma and Stompff 2023; Stompff et al. 2021). Overlapping issues in urban governance cannot

DOI: 10.4324/9781003449027-11

be solved without bringing together multiple actors and domains of knowledge that don't readily collaborate – necessitating new mechanisms for social innovation in the built environment (Horgan and Dimitrijević, 2018). In recent years, the living lab has emerged as a platform for joint problematisation of wicked problems – as a way to develop and test new solutions with the explicit participation of the community and cross-sectoral stakeholders (Koens et al. in press). As a step toward more genuine participatory governance, the living lab has shown great potential, in bringing stakeholders together and boosting the ability for collaboration across a tourism ecosystem (Horgan & Koens 2024). The need to move beyond siloed working that focuses on niche impacts, as opposed to fundamental societal drivers of wicked problems, necessitates new partnerships and ways of working. While destination marketing organisation are increasingly moving towards ecosystem management, the education system needs to prepare the practitioners of the future to engage in innovative design-led forms of collaboration. In response, and during the uncertain period caused by the pandemic, Inholland University of Applied Sciences took a daring step to foreground the living lab model more explicitly, in an adapted version and accessible format for various study programmes in their education offer. By doing so, it was able to bring tourism, leisure and other students out of the classroom and into the real world together, or via online possibilities when needed due to COVID-19 (Inholland University of Applied Sciences 2023).

The development of the Inholland Urban Leisure and Tourism (ULT) labs coincides with a reimaging of the tourism offer in the Netherlands – in support of local communities and the needs of actors across the entire destination ecosystem. In Amsterdam for example, spreading tourism to peripheral communities (multi-core city development) is a central part of the urban governance strategy (Gemeente Amsterdam 2022). In practice, however, this possibly brings with it new socio-spatial conflicts, like gentrification and a deficit of local ownership over development policies. It is within this unique context of urban transformation that living labs thrive, and this is a reason for the location of the Urban Leisure and Tourism labs in both Rotterdam and Amsterdam. Issues in Amsterdam (and the desire of municipalities such as Rotterdam not to face a similar fate) have opened up a space for the participation of a quintuple helix set of actors that together can unpack high-level local challenges in destination management. In both cities, the lab works closely with policymakers from the local authorities, the local DMOs, activists and (social) entrepreneurs. Those involved are keen to avoid stimulating overtourism and instead wish to develop a sustainable tourism approach focused on neighbourhood diversity and positive tourism. Like in Rotterdam, the '*do-rism*' concept emphasises the importance of quality of life and place not only for visitors but also for local communities. While Rotterdam faces a different set of tourism challenges to Amsterdam, the focus on improving the quality of life and place for both locals and visitors is of prime importance. As a forum for imagining future scenarios, the labs create space for learners to co-imagine interventions that embrace 'localhood'.

This chapter will explain how through the labs, students are encouraged to develop and apply new tools and pathways for co-design and co-production with these actors, eventually arriving at impactful interventions that position tourism as a positive phenomenon in society. As a site of experimentation, the ULT labs seek to identify and test tourism and leisure interventions that improve the experience for both locals and visitors alike (Koens 2021). In tourism planning, interventions can be understood as purposeful place-based actions carried out by actors within a tourism ecosystem seeking to impact or alter a present scenario (Moretti 2021; Matteucci et al. 2022). In considering how interventions are conceived of and rolled out, this chapter focuses on the role of the living lab, as a learning environment for a whole destination ecosystem. Learners are not only our students, but likewise, several involved stakeholders experience how to surf the emergent waves of an ever-changing context. As for the learning directors who stimulate the design thinking climbs at our labs,

> Working in a lab can be exciting and rewarding, but it can also be challenging and requires hard work. Particularly when things may not be clear ... The work environment can feel intense and requires a lot of attention and motivation. However, with the correct support, you will be able to experiment and create.

Zac Woolfitt shares the above reflection, alongside other lab actors in an upcoming publication on learning in living labs (Desomviele et al. 2024).

Urban living labs as a tool for regenerative placemaking: Towards social innovation in the built environment

For EnoLL (European Network of Living Labs), living labs are real-life test and experimentation environments that foster co-creation and open innovation among the main actors of the quadruple helix model. Their conceptualisation builds on the working definition of urban living labs used by the Joint Programming Initiative (JPI) Urban Europe up until 2019:

> a forum for innovation, applied to the development of new products, systems, services, and processes in an urban area; employing working methods to integrate people into the entire development process as users and co-creators to explore, examine, experiment, test and evaluate new ideas, scenarios, processes, systems, concepts and creative solutions in complex and everyday contexts.
>
> (JPI Urban Europe 2015: 59)

While it is beyond the scope of this chapter to offer a full overview of the academic debate on the different classifications of (urban) living labs that exist (for this, see e.g., Hossain et al. 2019; Leal Filho et al. 2023), it is useful to differentiate (urban) living labs from other types of lab environments, as a social lab.

Hassan (2014) first described the phenomena of tackling complex problems through the explicit participation of local stakeholders and community actors, in 'social labs'. They are conceptually different from other types of labs due to the devolution of power, participation and governance that they encourage – away from top-down or paternalistic projects that exist within the municipality or local government setting. It is the plurality of participation that sets them aside from other, less open models. They are characterised as place-based, and accepted and embedded within their local context with an eye to increasing sustainability and/or developing capacities for resilience. Understood as social labs, the first living lab in Urban Leisure and Tourism was founded in 2015 in Amsterdam, preceding the Rotterdam lab which opened during the pandemic in 2020. In both cities, the labs are located in peripheral city districts undergoing significant socio-spatial transformation, and facing related economic, social and cultural challenges. In order for tourism to contribute to social innovation in the local context, students become partners in capacity building, helping to co-develop and inform lab practices through a fluid exchange with lab leads and learning coaches. The decision to embed applied education as a physical presence in those areas – and become part of the local ecosystem – called for a redesign in the way of educating students, whilst considering the city itself as the basis of a new curriculum.

In the tourism area, placemaking and regenerative development are two compatible approaches that centre on unique local attributes as part of a change process. Both are people- and place-centric, understanding settlements as complex socio-ecological ecosystems (Hernandez-Santin et al. 2020). While there are conceptual differences between them, these are complementary. Through their combination, they offer a model for holistic, systemic change in tourism towards sustainable, locally owned interventions. Regenerative placemaking has thus emerged as a guiding theoretical basis for connecting our research and practice in the living labs, marrying concepts of regenerative development with place-based processes. While imbued with regenerative principles, key characteristics of placemaking – deep community engagement, small interventions to trigger long-term benefits, and improving quality of life by place attachment and resilience – are emphasised (Kyle et al. 2004; Hernandez-Santin et al. 2020).

The model proposed by Hernandez-Santin et al. (2020) understands placemaking as a temporal activity, nested within regenerative development pathways, but allows the potential of the place to reveal and manifest itself. The model emphasises regenerative placemaking as iterative practice – a creative way of trialling key community-led interventions related to long-term development outcomes. While we are actively involved with international placemaking networks, and organisations such as STIPO and European Placemaking Network – and share various known placemaking models with our students – we are cognisant of community criticism of *placemaking*. We noticed that the word placemaking can provoke connotations of stimulating gentrification and social exclusion – particularly in dialogue with some local activists who see the practice as adjacent to neoliberal urban governance. While we recognise these perspectives, within the lab community it underlines the complexity for our learners to really co-design from a multi-stakeholder approach. While *regenerative placemaking* is still orienting practice away from top-down approaches to urban renewal, it can lack robust evaluated examples of social transformation. In instruction therefore, lab leads tend to describe what we are aiming for in terms of impact on end beneficiaries: how we work and show concrete examples that have been developed using our methods. Across the practitioner landscape, there is consensus around the need to better 'translate' regenerative design concepts and amplify their potential as place-based interventions.

In seeking to develop durable and nuanced relationships with local partners and networks, regenerative placemaking marks a shift away from outmoded narratives and perceptions about place – a process that can rapidly catalyse engagement, and cultivate empathy and local ownership over change (Hernandez-Santin et al. 2020). In the years it has been active in Amsterdam Noord, the lab has built meaningful relationships with amsterdam&partners (destination management organisation), the municipality, shopping centre Boven 't Y Winkelcentrum, sailing Tours that Matter, Cinekid, cultural hub Modestraat and many others. These stakeholders serve as project sponsors, helping students to navigate the local context – sharing their tacit knowledge of a situation to plant the seeds that encourage a cross-pollination of ideas, for which a lively learner population is a vital ingredient. Assisting with brokerage into the local community as a strategy towards widening engagement, participation and visibility are therefore core to the work of the lab (Jernsand 2019). Inholland's unique urban leisure and tourism living labs use themes such as regenerative placemaking to guide students, using design thinking to develop responsive propositions that can be devolved to local entrepreneurs and changemakers (Koens et al. in press). Within tourism, the need for reflective practices is increasingly argued for, particularly in the context of resilient and regenerative tourism practices (Bellato et al. 2023; Bellato and Cheer 2021; Koens et al. 2021). In order to embed this way of working into our learning community, a set of bespoke lab tools was developed. The Place Exploration and Sense Making Map (Gerritsma 2021) helps to reveal local tacit knowledge about the place, and the Co-design Canvas was intentionally adapted for use in lab settings to widen participation and open evaluation (Smeenk and Bertrand 2020).

Making and breaking tourism interventions: Labs as learning environments for iterative experimentation

In this section, we share key reflections on the development of a robust stakeholder ecosystem and building a community of practice, based on our experiences in our Urban Leisure and Tourism labs in Rotterdam-Zuid and Amsterdam-Noord. Since 2020, both labs have accommodated a semester-long programme (similar to a minor course) in which third-year students learn to apply principles of design in a real-world environment, in response to a challenge the lab leads bring in, based upon their research agenda and in close contact with the local ecosystem. Students work largely with models that understand design as a reflective practice – and are encouraged to take a critical perspective and infuse their learning approach with social design principles. For this, the labs work closely with residents, municipalities, and industry as well as grass-roots organisations. Around each challenge a different set of stakeholders coalesce, helping students understand the local context, verify assumptions and develop coherent design propositions (e.g., align ideas to broader urban strategies, policy agendas and business ventures). The set-up is very different from a traditional curriculum in tourism. It is intended to provide structure for students who are not used to the iterative nature of design approaches and is organised around four design sprints. These are referred to as 'climbs', in order to emphasise the effort required. After each sprint the students are required to present ideas to the external stakeholders, generating insights and immediate feedback.

In forming responsible young professionals, students are stimulated to act as changemakers and *design with the community*, not for them. This is not without difficulty, as some students are not comfortable in engaging with communities in difficult conversations about equity, while other young people can hold engrained (populist) biases towards sections of society in need. While the structure of the climbs helps students understand the process of design thinking in developing concepts that can have eventual local ownership, it does take away some of the flexibility and open-mindedness that exemplify living lab work – the spontaneity that can sometimes encourage radical ideas. To support and guide our students, a special Inclusive Design Toolbox has been developed. This can help integrate the ideas of inclusive design into the curriculum as much as possible and sow seeds where possible, through responsive interventions (Collin & Gerritsma 2022). In addition, various workshops and trainings are incorporated within the curriculum, to embed inclusive design practice and raise awareness by highlighting impactful interventions. It is not just students who have to adjust to design-based learning. Lecturers and facilitators at universities are often inexperienced on this matter as well and can have difficulty themselves adjusting to this kind of learning. These issues can stifle momentum and lead to dissatisfaction among students.

Based upon our semestrial evaluations, it turns out that after a few iterations, the value of the living labs is not always directly fully understood by many students, even though each semester a small group is highly engaged and enthusiastic about design-led education in a lab. When it comes to the outcomes of lab-based education, students agree on the value of learning from real-life experiences, in comparison to other courses. Also, overall, they very much enjoy the creative nature of design thinking and the fact that they are challenged to think about different perspectives on, for instance, sustainability and social equity. Empowering students through co-creation processes also creates new uncertainties as to the outcome of courses. Within the current course structure, reflective practices focus strongly on personal development – in line with the rest of the curriculum – and could be enriched with reflections on external developments. This is one of the reasons why the labs are weaving the co-design canvas more deeply into the adjusted learning journey, to stimulate collective reflection from various stakeholder perspectives as well (Smeenk & Bertrand 2020). Within both learning communities in Amsterdam and Rotterdam, students have designed various concepts with the aim of building more attractive places for all city users – be these residents or tourists, entrepreneurs or day-trippers – in line with the prevailing discourse on sustainable urban tourism (Aall & Koens 2019). These range from interventions that seek to improve the quality of life and place, initiate new relationships between locals and visitors, and stimulate opportunities for positive tourism.

One such example is the 'Inclusive Welcome Path', proposed by students in Amsterdam after having conducted forensic mapping of the local area – using the Place Exploration and Sensemaking tool, observations, and interviews. This group understood that some people don't feel at home or welcome in the Buikslotermeerplein area – which they characterised as grey, unsafe, and boring. The concept of the Inclusive Welcome Path received very positive feedback from local stakeholders – who had direct input in creating the paper prototype. Unlike a standard rainbow path, the intervention is designed to be welcoming for everyone and was immediately embraced by citizens during prototyping, who spontaneously added the word for 'Welcome' in their own languages. Local capacity has the power to make or break a concept, as in the case of this intervention a finished product wasn't realised, despite implicit support by the local municipality. On this occasion, more time is needed and idea owners lack the capacities to take the idea to fruition or dedicated people to take it further. However, a good idea might fall on fertile soil later on. The lack of collective ownership and the inability to actually implement proposed interventions remains a barrier to the adoption of these ideas, as in the case of another concept to originate in the Amsterdam lab related to urban greening. In this case, the concept was not taken on board locally due to a similar lack of local stakeholders who could maintain it. This project revealed the challenges that exist around ownership and biodiversity in the public realm – and allowed both research and practice to engage with the complexity of the delivery and implementation of social innovations. Similarly in Rotterdam, another concept to emerge related to local policies of upgrading parks was the BMX-route Zuidpleinen. That concept, which introduced BMX routes to ferry cyclists between parks in Rotterdam Zuid, included a number of valuable concepts to address socio-spatial inequality in those neighbourhoods. Ultimately what makes or breaks these interventions in practice is the ability to infuse them with local governance, ownership, and participation (Horgan & Dimitrijević 2020).

In our labs, the process of making and breaking tourism interventions is a continuous one, which recognises the dynamic nature of co-production – which is not without its own challenges. Our vision for the next stage of interventions being developed after a lab track within the living labs is more akin to a production house where less-developed concepts may be taken off the shelf and trialled (and evaluated) with communities in situ.

Discussion and conclusion: (Regenerative) placemaking as iterative experimentation – a process of breaking and resetting

Collaborative design is a messy and complex set of activities without a definite point of origin or arrival, which can often be frustrating for those unaccustomed to uncertainty. While designers are taught to embrace this ambiguity, we often underestimate how difficult it is for students from other disciplines to work in such a nuanced way, outside of a traditional pedagogical environment. In essence, design thinking methodologies must be learnt by doing, meaning learners need to develop skills in self-actualisation and critical reflection. Summarising the experience of initiating urban living labs for leisure and tourism, we offer a number of considerations for what can *make or break* tourism interventions in the lab setting.

What can make an intervention:

- Holding space – as a rather neutral actor the living lab represents the connecting tissue or glue that brings the community together.
- Fountain of inspiration – the lab ecosystem must provide energy and resources to revitalise and seed ideas to flourish.
- Long-term presence and long-term commitment projects – building trust by staying in the community for years, actively demonstrating a legacy of experimentation and spending time for follow-ups of potentially rich concepts.
- Reflection as part of our way of working – we take a step back to create space for feedback loops with end beneficiaries and reflective co-learning across the ecosystem. Every lab track ends with a formal evaluation among students and local partners. These feedback loops are very insightful and have led to several adjustments for the following lab tracks. Informal checks are carried out constantly and are considered when possible.

What can break an intervention:

- Getting stuck in ideation – the lab must push stakeholders to move past conceptualisation to real-world testing and evaluation.
- Lack of local ownership – need to devolve responsibilities to local actors, collective ownership, and duty of care to interventions.
- Resentment against the intervention – a robust sense-making phase about and sensitivity towards the place and local community are both delicate and vital.

Above all, innovation in the living lab setting calls for individuals who dare to fail or fail forward and embrace uncertainty, yet are committed to navigating a path towards impact. This in essence represents making and breaking as iterative prototyping; breaking established systems and orthodoxies. Research by design sits well within this context with the aspiration to create lasting, meaningful, real-world impact. The specific challenges highlighted in this chapter emerge from the growing pains felt by our labs.

As we have learned over the development of our living labs, (regenerative) placemaking (although not always calling it as such) provides a meaningful framework for guiding the complexities of urban transformation – and provides a platform for expanding the role of new urban tourism and leisure to support social innovation and spatial inclusion (Horgan 2019).

While it is important to be honest about the difficulties encountered with scaling concepts, there are others that were immediately embraced by locals – due in part perhaps to their simplicity. Successful examples in this sense are, for example: a 3×3 basketball court and the Verhalenbank in Amsterdam Noord, both in close collaboration with various partners. A Rotterdam project, in collaboration with local organisation Stichting Werkshop Foundation, involving a tour for asylum seekers with granted refugee status, has also been incredibly well-received. Moving forward, we understand, based upon feedback from our local partners, the need for our labs to push more often beyond ideation and conceptualisation with communities to steward the development of viable interventions that, through testing, build local ownership and capacity for implementation. Bylund et al. (2022) note that as part of a wider politics of experimentation, in which the governance of urban sustainability is being negotiated, ULT labs need to be better integrated in order to drive larger paradigm shifts and tackle democratic deficits (Karvonen 2018).

In terms of interventions having the capacity to impactfully make or break a situation for the better and in the long run, we are therefore exploring possibilities of adding another layer to the labs as production houses, where off-the-shelf student concepts can be taken on and experimented with by other stakeholders within our ecosystems – entrepreneurs, companies, and public sector entities. Our alignment with the EnoLL network is a step in the right direction and will help us to engage in collective thinking – around permanence versus temporality in interventions, the duty of care to local communities and how better to integrate innovative versus traditional instruction-led education.

Equally, as we embrace systems thinking and regenerative practices, our labs can become focal points to unpack the critical connotations associated with placemaking. Resilience, regeneration and development can often be considered loaded terms – reminiscent of social engineering, where the ambitions of development actors are assumed, and where place-based actors at the grass-roots lack effective agency. Indeed, as pushback grows around the temporality of placemaking activities and new tourism offers in general, we must continue to self-reflect and question our role in development, and that of tourism as a social practice.

Within the realm of regenerative design, metaphors related to nature are frequently used. In order for tourism interventions to break through, and catalyse lasting change, we increasingly call ourselves urban gardeners, working our way around to allowing green shoots to penetrate rigid systems. Making or breaking an intervention turns out to be a thin red line, as often 'the proof of the pudding is in the eating' and depends on how well an intervention is falling on fertile soil and is embedded via the lab by the local community.

References

Aall, C., & Koens, K. (2019). The discourse on sustainable urban tourism: The need for discussing more than overtourism. *Sustainability*, *11*(15), 4228.

Bellato, L., & Cheer, J. M. (2021). Inclusive and regenerative urban tourism: Capacity development perspectives, *International Journal of Tourism Cities*, *7*(4), 943–961.

Bellato, L., Frantzeskaki, N. & Nygaard, C. A. (2023). Regenerative tourism: A conceptual framework leveraging theory and practice. *Tourism Geographies*, *25*(4), 1026–1046.

Bylund, J., Riegler, J. & Wrangsten, C. (2022). Anticipating experimentation as the 'the new normal' through urban living labs 2.0: Lessons learnt by JPI Urban Europe. *Urban Transformations*, *4*(1), 1–10.

Collin, P., & Gerritsma, R. (2022). Inclusief Leisure Ontwerp in het hbo-onderwijs: Een meanderend pad. *Vrijetijdstudies Jaargang*, 40, 27–38.

Desomviele, L., Vandevyvere, I., van den Hee, M. & Woolfitt, Z. (2024, in press). *Learning in a living lab: Knowing what to do when you don't know what to do.* Ghent: Owl Press.

Enterprise Network for Sustainable Urban Tourism (2023). New Urban Tourism: Magazine for the critical friend and urban explorer. Available from: https://www.ensut.eu/wp-content/uploads/sites/11/2023/02/NUT_magazine_ENGLISH_v25022023.pdf (Accessed 12 October 2023).

Gemeente Amsterdam (2022). Visie bezoekerseconomie Amsterdam 2035. Available from: https://openresearch.amsterdam/nl/page/90775/visie-bezoekerseconomie-amsterdam-2035 (Accessed 17 February 2024).

Gerritsma, R. (2019). Overcrowded Amsterdam: Striving for a balance between trade, tolerance and tourism. In: Milano, C., Cheer, J. M. & Novelli, M. (Eds.), *Overtourism: Excesses, discontents and measures in travel and tourism*. Abingdon: Cabi.

Gerritsma, R. (2021). Place Exploration and Sense-Making Map. Available from: https://www.ensut.eu/knowledge/place-exploratation-sense-making-tool/ (Accessed 17 February 2024).

Gerritsma, R., & Stompff, G. (2023). Who is at Amsterdam's tourism policy-making table? In: Hospers. G-.J. & Amrhein, S. (Eds.), *Coping with overtourism in post-pandemic Europe*. Münster: LIT Verlag, pp. 116–130.

Hassan, Z. (2014). *The social labs revolution: A new approach to solving our most complex challenges*. Oakland, CA: Berrett-Koehler Publishers.

Hernandez-Santin, C., Hes, D., Beer, T. & Lo, L. (2020). Regenerative placemaking: Creating a new model for place development by bringing together regenerative and place-making processes. *Designing Sustainable Cities*, 53–68.

Higgins-Desbiolles, F., & Everingham, P. (2022). Degrowth in tourism: Advocacy for thriving not diminishment. *Tourism Recreation Research*, *49*(1), 1–5.

Horgan, D. (2019). Placemaking. In: Thrift, N. & Kitchin, R (Eds.), *International encyclopedia of human geography*. Amsterdam: Elsevier, pp. 145–152.

Horgan, D., & Dimitrijević, B. (2018). Social innovation systems for building resilient communities. *Urban Science*, *2*(1), 13.

Horgan, D., & Dimitrijević, B. (2020). Social innovation in the built environment: The challenges presented by the politics of space. *Urban Science*, *5*(1), 1.

Horgan, D., & Gerritsma, R. (2023). Sense and spreadability: Considering how data and visibility could support more engagement-led socio-spatially inclusive pride events in peripheral neighbourhoods of Amsterdam. *Tourism Cases*, (2023), tourism202300404.

Horgan, D., & Koens, K. (2024). Don't write cheques you cannot cash: Challenges and struggles with participatory governance. In: H. Pechlaner, E. Innerhofer & J. Philipp (Eds.), *Overtourism to sustainable governance: a new tourism era.* London: Routledge.

Hossain, M., Leminen, S. & Westerlund, M. (2019). A systematic review of living lab literature. *Journal of Cleaner Production*, *213*, 976–988.

Huijbens, E. H., & Jóhannesson, G. T. (2020). Urban tourism. In: Jensen, O. B. (Ed.), *Handbook of urban mobilities*. London: Routledge, pp. 314–324.

Inholland University of Applied Sciences (2023). Living labs. Available from: https://www.inholland.nl/onderzoek/lectoraten/studiesucces/living-labs/ (Accessed 30 August 2023).

Jernsand, E. M. (2019). Student living labs as innovation arenas for sustainable tourism. *Tourism Recreation Research*, *44*(3), 337–347.

JPI Urban Europe (2015). Strategic research and innovation agenda: Transition towards sustainable and liveable urban futures. Available from https://www.jpiurbaneurope.eu/

app/uploads/2016/09/JPI-UE-Strategic-Research-and-Innovation-Agenda-SRIA.pdf (Accessed 18 February 2024).

Karvonen A. (2018). The city of permanent experiments? In: Turnheim, B., Kivimaa, P. & Berkhout, F. (Eds.), *Innovating climate governance: Moving beyond experiments.* Cambridge: Cambridge University Press, pp. 201–215.

Koens, K. (2021). *Inaugural address on reframing urban tourism.* Available from: https://www.inholland.nl/nieuws/en/ko-koens-delivers-inaugural-address-on-reframing-urban-tourism/ (Accessed 18 February 2024).

Koens, K., & Milano, C. (2023). Urban tourism studies: An integrative research agenda. *Tourism, Culture and Communication*, DOI:10.3727/109830423X16999785101653.

Koens, K., Postma, A. & Papp, B. (2018). Is overtourism overused? Understanding the impact of tourism in a city context. *Sustainability, 10*(12), 4384.

Koens, K., Smit, B. & Melissen, F. (2021). Designing destinations for good: Using design roadmapping to support pro-active destination development. *Annals of Tourism Research, 89*, 103233.

Koens, K., Stompff, G., Vervloed, J., Gerritsma, R & Horgan, D. (in press). How deep is your lab? Understanding possibilities and limitations of living labs in tourism. *Journal of Destination Marketing and Management.*

Kyle, G., Graefe, A., Manning, R. & Bacon, J. (2004). Effect of activity involvement and place attachment on recreationists' perceptions of setting density. *Journal of Leisure Research, 36*(2), 209–231.

Leal Filho, W., Ozuyar, P. G., Dinis, M. A. P., Azul, A. M., Alvarez, M. G., da Silva Neiva, S., et al. (2023). Living labs in the context of the UN sustainable development goals: State of the art. *Sustainability Science, 18*(3), 1163–1179.

Lew, A. A. (2019). Tourism planning and place making: Place-making or placemaking? *Tourism Geographies, 19*(3), 1–19

Matteucci, X., Koens, K., Calvi, L. & Moretti, S. (2022). Envisioning the futures of cultural tourism. *Futures, 142*, 103013.

Moretti, S. (2021). State of the art of cultural tourism interventions. Deliverable D3.1 of the Horizon 2020 project SmartCulTour (GA number 870708). Published on the project web-site in May 2020: http://www.smartcultour.eu/deliverables/ (Accessed 26 February 2024).

Oskam, J. A., & Wiegerink, K. (2020). The unhospitable city: Residents' reactions to tourism growth in Amsterdam. In: Oskam, J. A. (Ed.), *The overtourism debate: NIMBY, nuisance, commodification* (pp. 95–118). Bingley: Emerald Publishing Limited.

Rotterdam Make it Happen (n.d.). the Dorist. A brand-new expression for a different kind of tourist. Available from: https://www.dorotterdam.com/the-dorist (Accessed 17 February 2024).

Smeenk, W., & Bertrand, G.(2020). Het Co-design Canvas: Een empathisch co-design instrument met maatschappelijke impact. Author report.

Stompff, G., Gerritsma, R., Waterreus, S. & Koens, K. (2021). *Amsterdam case study report.* SMARTDEST H2020. Amsterdam: Inholland University of Applied Sciences.

Tourismlab Amsterdam (2022). Studenten ontwerpen 3x3 basketbalveld voor community op Buikslotermeerplein. Available from: https://www.tourismlabamsterdam.nl/2022/01/14/tudenten-ontwerpen-3x3-basketbalveld-voor-community-op-buikslotermeerplein/ (Accessed 31 August 2023).

Tourismlab Amsterdam (2023). Het Verhalenbankje in het nieuws. Available from: https://www.tourismlabamsterdam.nl/publications/het-verhalenbankje-in-het-nieuws/ (Accessed 31 August 2023).

Urry, J. (2016). *What is the future?* Chichester: John Wiley & Sons.

Urban Leisure and Tourism Lab, Amsterdam (2023). Het verhalenbankje in het nieuws. Available from: https://www.tourismlabamsterdam.nl/publications/het-verhalenbankje-in-het-nieuws/ (Accessed 31 August 2023).

Urban Leisure and Tourism Lab, Rotterdam (2023). A Tour of Hart van Zuid. Available from: https://www.tourismlabrotterdam.nl/en/activities/een-tour-door-hart-van-zuid/ (Accessed 31 August 2023).

10 The Role of Participatory Approaches in Developing Future Perspectives

Reconversion of St Godelina's Abbey, Bruges

Bart Neuts, Steven Valcke, Clio Lambrechts, Vincent Nijs, and Jan Van Praet

Introduction

During the early stages of commercial tourism in the 1950s, when international tourism was still in its infancy, growing visitor numbers were seen as desirable due to the economic benefits they could bring to a destination. Following the destructive forces of World War II, international tourism arose thanks to new applications of the technological progress that was made, particularly in mobility, as well as the progress in social laws and growth in disposable income, which offered opportunities for leisure time. Also from an international, governmental perspective, there was a vested interest in cross-border collaborations, exemplified by increased globalization. Even though the earliest national tourist organization (NTO), tasked with supporting tourism development at the national level, was already established in New Zealand in 1901, Borzyszkowski (2015) notes how the vast majority of NTOs were created in the early post-war period. NTOs are also commonly referred to as destination management organizations, or destination marketing organizations (DMOs), the difference being that the term NTO explicitly refers to the national context, while DMOs can exist on multiple geographical levels. In the continuation of the chapter, we will use the term DMO as the more general terminology.

In her critical analysis of the historic and contemporary role of DMOs, Dredge (2016) discusses how a typically industrial ontology underpins the values of DMOs, with the organizations often being a policy tool to manage industry interests and chiefly being centred around tourism marketing – and to a lesser degree, development of tourism products – to leverage the economic opportunities provided by the growing sector. However, as noted by multiple authors (e.g., Wang 2011; Woodside & Sakai 2009), the role and functioning of DMOs have evolved significantly over the four decades of near continuous tourism growth. These changes have partly been informed by a growing understanding of the possible adverse social, cultural, and environmental impacts of tourism which have received significant scientific attention since the early 1990s. While earlier research had predominantly focused on environmental effects, as well as socio-cultural shifts in host communities and indigenous peoples, the "overtourism" debate, which gained much traction in

DOI: 10.4324/9781003449027-12

Western European cultural and urban destinations since around 2015, was a clear indication of the limits to growth not just in nature-based and less-developed regions of the world, but also in urbanized settings.

As a result, various authors have called upon DMOs to increasingly pivot from place marketing to place management as a policy priority, embracing social responsibility towards both visitors and locals, stewardship of local culture and resources, and sustainability (Morgan 2012; Morrison 2013). Pechlaner et al. (2019) specifically identify eight functions of contemporary DMOs: (i) destination marketing, branding, and positioning; (ii) relationship building/coordination; (iii) product development/development activities; (iv) destination planning, strategy formulation, monitoring and evaluation; (v) resource stewardship; (vi) human resource development; (vii) visitor management; and (viii) enhancing well-being of destination residents. For many DMOs, the broader responsibilities of destination stewardship and a shifting perspective from quantity to quality have significantly changed both strategies and actions. The involvement of the local society has come to play an important role, since informed and involved residents often have a more positive perception of tourism, leading to enhanced well-being of a destination's residents (Jainchill n.d., Šegota et al. 2017).

In this chapter, we introduce a pilot study of the DMO of the Dutch-speaking Flemish region of Belgium, VISITFLANDERS, and their involvement in a revaluation project of St Godelina's Abbey in Bruges. The case study can be insightful since VISITFLANDERS has followed a strategic and organizational repositioning in line with modern stewardship approaches, reimagining tourism as a means to local well-being and thereby placing the local stakeholders central in the tourism development discourse. As part of the strategic operationalization, participatory approaches are followed in larger-scale intervention projects such as the purchase and subsequent reconversion of St Godelina's Abbey. In the remainder of the chapter, first of all, the vision, strategy, and adopted changes of the organization are discussed, as well as the brief history of St Godelina's Abbey in order to frame the case study. Next, we offer details on the research methodology that was followed for the participatory inquiry, as well as the subsequent analysis of results, and a technical evaluation. The results should offer insights into practical implementations of local involvement, and its potential benefits and limitations.

Case study: VISITFLANDERS and St Godelina's Abbey

While the first Belgian National Bureau for Tourism was already established in 1921, the first real NTO in Belgium was the Commissariat-General for Tourism in 1939. Due to the political structure of Belgium, with French-speaking, German-speaking, and Dutch-speaking communities, the national tourism organization was split into three autonomous administrations in 1982, as part of extensive state reforms. Therefore, VISITFLANDERS – albeit until 1995 under a different name – has been responsible for tourism promotion and development of the Dutch-speaking region of Belgium since 1982 (De Groote 1999). Compared to many other DMOs, it is worth noting that VISITFLANDERS has had relatively broad responsibilities throughout the past, combining the promotion of Flanders internationally with product development at the destination – both directly and via subsidy programmes, quality control and regulation, and social tourism. Even though the organization has therefore never been purely marketing-based, traditional KPIs were still partly measured in arrivals, overnight stays, and economic contributions.

In 2017, VISITFLANDERS embarked on a co-creative process, together with the broad tourism sector, visitors, local residents, and national and international experts to reflect on the role of tourism in a changing and evolving society. The resulting research synthesis was published in 2018 and entitled "Travel to Tomorrow", which included a reconceptualization of tourism as a means, rather than a goal, with the end goal being to support flourishing destinations – including their residents, visitors, and local entrepreneurs. Tourism can be a positive power, as long as it is managed sustainably and with respect to local carrying capacities and priorities. The new concept inspired the policy agenda for 2019–2024 and was further operationalized in the strategic document "Flora et Labora", which proposes five core tenets: (i) creating an added value for all stakeholders: residents, visitors, entrepreneurs, and the place itself; (ii) connecting people, places, and activities through a unique story; (iii) participation of stakeholders; (iv) qualitative experiences with room for innovation and creativity; and (v) knowledge-driven operations. These principles run through six destination themes – heritage experience, natural Flanders, culinary Flanders, Flanders cycling, conferences and events, social tourism – and are applied and tested in so-called pilot projects (VISITFLANDERS 2021).

One such pilot project is St Godelina's Abbey in Bruges. The abbey, located just within the historic centre of the city, has been a listed protected monument since 1991. The history of St Godelina's Abbey can be traced back to 1586, when the Benedictine sisters of St Godelina of Gistel sought refuge within the city of Bruges. The first physical structures date to 1623, when a first house and yard were bought to house a number of sisters, which was followed three years later by the construction of a church and choir. From 1626 to 1645 multiple additions were made through the purchase of surrounding properties and the (re)construction of buildings, including a refectory, kitchens, and brewery. During the next century, the focus was mainly on reparations and restorations while 1885 saw the construction of a new wing. The most recent constructions date to 1953 when a final wing was built at the east side of the property (Inventaris Onroerend Erfgoed 2023). The abbey served its religious purpose until 2013, when the last Benedictines left the site, and functioned as a closed convent in which residents had little to no contact with the outside world and were largely self-sufficient. After having been curated by engaged local volunteers of the Brugge Foundation for a few years, VISITFLANDERS acquired the property in leasehold in 2021 with the objective of reappraising and disclosing the heritage to locals, visitors, and tourists. The case of St Godelina's Abbey is interesting since it connects with a problem that is more commonly experienced throughout Western Europe as a result of increased secularization: how to safeguard and repurpose valuable religious heritage sites after they have lost their original religious function?

Research Methodology

Since the strategic vision of VISITFLANDERS is to leverage tourism as a positive means to support flourishing communities – benefiting the quality of life of its local residents as well as contributing to a qualitative visitor experience – a core aspect of the operationalization relates to participatory co-design approaches. St Godelina's Abbey functioned as a pilot project in designing a large-scale qualitative referendum to collect local perceptions and ideas on future development initiatives.

In August and September 2021, the abbey was opened to the general public for the first time in nearly 400 years, in the original state it was in since the last of the Benedictines left the site. Visitors followed a predefined route which included a general introduction and a few experiential elements, such as seeing ongoing architectural renovation works or eating a snack in the refectory. During this period, visitors were invited to contribute their ideas for future development scenarios. Individual scenarios were collected from the site visitors via a structured paper questionnaire. More than 16,000 visitors were welcomed in this period and invited by VISITFLANDERS' staff members to participate in a survey. The questionnaires were respondent-completed and combined a larger open-ended question with five additional categorical and semantic differential scale-type questions. Table 10.1 provides an overview of the questions and question types. While a majority of visitors completed (part of) the questionnaire, due to the unexpectedly large attendance, not all responses could be analysed. In the end, the choice was made to only analyse the responses in cases where respondents had completed the open-ended question and the cases where respondents did not merely prefer the status quo. Ultimately, 4,784 responses were analysed, 51.6% coming from Bruges or a neighbouring municipality and 48.4% being excursionists and day visitors. To ensure that there was no self-selection bias present, at a later date, the results of 211 additional questionnaires (which were not originally analysed due to a missing open-ended future scenario) were compared with the results of the original selection. This comparison indicated no significant differences between the original sample (n=4,784) and the test sample (n=211).

Results

When first analysing the open-ended main question on the future vision for St Godelina's Abbey's reconversion, a narrative analysis identified key terms and ideas. The proposed ideas of respondents were typically layered and multidimensional, not just focusing on one single aspect of the site. A few themes were most prevalent throughout many narratives, namely reference to the abbey garden, which was often linked with qualities such as tranquillity, silence, relaxation, contemplation, and gardening, but there were also ideas regarding integrating the garden in locally embedded events and as a place for sharing experiences and supporting local entrepreneurship and artisanal production. The following two

Table 10.1 Composition of the individual scenario collection instrument

Question	Answer categories	Type
We are in 2025 … You take a domestic or foreign visitor to this place. What can they do and see here? What is happening at various locations? Who is walking around here? What can they hear, see, smell, taste, feel?	None (open-ended)	Open-ended
In your future scenario, which three activities can take place here concurrently? (Select maximum three options)	(1) eating and drinking, (2) exhibitions/ateliers, (3) reflection, (4) debates on societal issues, (5) performances (music, theatre, dance, …), (6) meeting people, (7) lodging and accommodation, (8) visit to (religious) heritage, (9) activities for children/youths, (10) activities for local associations, (11) living, (12) training and education, (13) temporary pop-up activities and events, (14) sports, (15) liturgy and church experience, (16) work, (17) other	Categorical
What atmosphere does this place exude in your future vision? (Mark your position on a scale between two extremes with an x)	Open and visible ⟷ Closed and hidden Rest and silence ⟷ Dynamic and bustling Light ⟷ Dark Modern ⟷ Historical Subdued ⟷ Expressive Clear ⟷ abbey atmosphere ⟷ No abbey atmosphere	Binary semantic differential scale
What is central in your future image? (Mark your position within the triangle with a cross. The closer you place your cross to one corner, the more your answer correlates with that word. If you place your cross in the middle of the triangle, all three words are seen as equally important)	The place / The people / The activities	Three-dimensional semantic differential scale
Whom is this place mostly for according to your future image? (Mark your position within the triangle with a cross. The closer you place your cross to one corner, the more your answer correlates with that word. If you place your cross in the middle of the triangle, all three words are seen as equally important)	Local entrepreneurs / Local residents / Visitors	Three-dimensional semantic differential scale
In your future image, this is a place for … (Mark your position on a scale between two extremes with an x)	Small no. of visitors ⟷ Large no. of visitors Visitors/tourists ⟷ Local residents Single dest./activity ⟷ Multiple dest./activities	Binary Semantic Differential scale

quotes are an example of such an integrated approach to a potential abbey garden reconversion:

> The garden could be used as an allotment and a quiet meeting place. On the other hand, space could be made for healthcare, which is accessible and where young healthcare practitioners can work on an independent basis. This is also an ideal place to realise the vision and research of life-me vzw of the VUB and do prevention within healthcare. (R1)
>
> The site was actually an example of how sustainable living was, before the term 'short chain' was used. I dream of redesigning the kitchen garden, orchard, farm … where produce is grown and processed in a cooking school with didactic restaurant. The students are newcomers to Bruges from all over the world. They introduce plenty of recipes from their countries of origin. But sustainability does not stop at food. A number of workshops are set up where 'old' Bruges residents can teach newcomers all kinds of crafts, crafts that are gone here. On the site, people will come and tell their stories, which will then be reflected upon. Stories from all corners of the world are passed on and serve as inspiration to be spiritually sustainable in life. (R2)

Other elements that were recurring in most future visions related to the historical and religious nature of the site, both in terms of architecture and in relation to the Benedictine sisters who used to live there. While contemporary interpretations for the site considered its potential for temporary (art) exhibitions and musical performances, it was mostly deemed essential to respect the heritage and support smaller-scale, local development initiatives, as indicated by the following future scenario:

> A cultural place for visitors and tourists with respect for history, heritage. Temporary exhibitions, guided tours of the abbey, giving local artists a forum, music performances, debates, and lectures. Possibly a place for contemplation for small groups with the possibility of staying overnight. No mass tourism. Selective and for people with a cultural purpose. Small-scale catering with organic and local products. Preserving the green character. (R3)

A final prevalent theme was similarly linked to the site's heritage, but predominantly focused on the social nature of the religious site, linked with words such as "care", "tolerance", "connections", "spirituality", and "resilience". Such future visions underlined the potential of St Godelina's Abbey to become a meeting place. This is reflected in the following visitor suggestions:

> This place has been a sacred place for 400 years because the soul of this place was shaped by the souls of the hundreds of women who lived and worked here in contact with, and inspired by, something sacred to them. Family, nature, beauty, caring for each other as sacredness today. A culture of caring should continue to mark this place. (R4)

International cultural centre for youth, relationship building and connection. Young people form the link between past and present. Ecological concept. Courses, trainings and retreats for young people in search of connection and their own limits. Connection and experience in the 21st century. (R5)

The analysis of the open-ended questions revealed relatively strong homogeneity on a more general level, indicating that the vast majority of people supported small-scale, locally embedded development initiatives, building on the existing historical atmosphere. Ideas seemed to gravitate towards "slow tourism"-style interventions, avoiding large-scale commercial developments in favour of more contemplative experiences.

To collect further, more detailed information, respondents were then asked to connect their future visions with potential on-site activities. The main activities that were considered by respondents were exhibitions/ateliers (46%), reflection (36%), visit to (religious) heritage (33%), performances (music, theatre, dance, ...) (33%), meeting people (24%), eating and drinking (22%), lodging and accommodation (14%), training and education (12%), and debates on societal issues (10%). The remaining activities were mentioned in less than 10% of cases. This ranking of priorities further indicates that the majority of respondents seem to prefer arts- and culture-based activities that have intrinsic links to the historical (religious) characteristics of St Godelina's Abbey. The semantic differential scale ranking of the proposed atmosphere of future interventions in St Godelina's Abbey, as seen in Figure 10.2, confirms the importance of the historical context in any new context, with "rest and silence", "historical", "subdued", and "clear abbey atmosphere" all clearly being preferred.

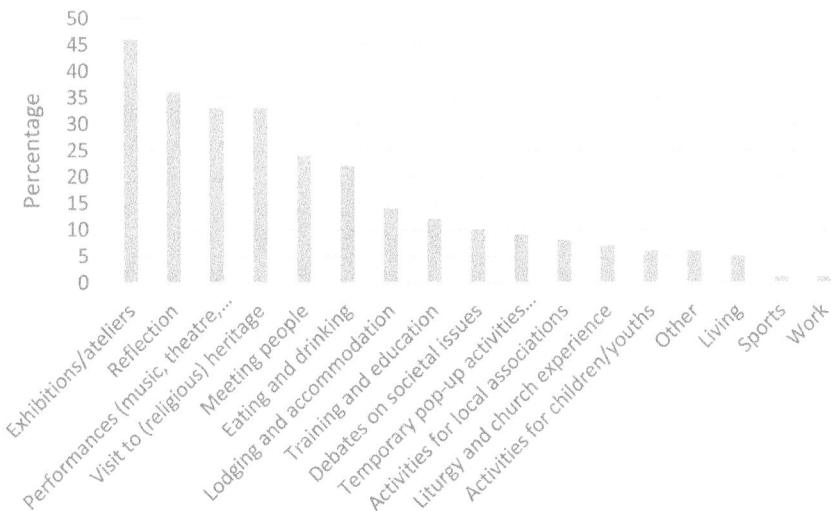

Figure 10.1 Potential future activities

Open and visible		**X**		Closed and hidden
Rest and silence	**X**			Dynamic and bustling
Light	**X**			Dark
Modern			**X**	Historical
Subdued	**X**			Expressive
Clear abbey atmosphere	**X**			No abbey atmosphere

Figure 10.2 Characteristics of a place (based on median values)

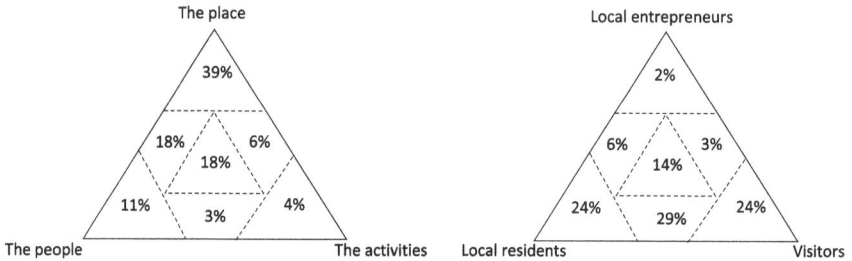

Figure 10.3 Priorities in future development scenarios

Further analysis of the three-dimensional semantic differential scale indicates priorities in concept and stakeholder inclusion. From the left side of Figure 10.3, we can identify that a majority (39%) consider the characteristics of place (atmosphere, history, architecture, etc.) as primary elements of future reconversion interventions. In general, respondents are more likely to locate themselves along the left side of the triangle (i.e., "place" – 39%, "place-people" – 18%, "people" – 11%) with less priority given to specific activities (4%). This seems to imply that social, architectural, and historical values ought to be respected and given prevalence over more commercialized developments that focus on activities without adequately integrating the socio-cultural reality. The right side of Figure 10.3 gives an indication of stakeholders who ought to be prioritized according to survey responses, again, reflecting less commercial incentives ("local entrepreneurs" – 2%), with answers converging around the base of the triangle (i.e., "local residents" – 24%, "local residents-visitors" – 29%, "visitors" – 24%). It should be mentioned though that these results might be influenced by the composition of survey respondents, with residents and visitors being more or less equally represented, and local entrepreneurs not having strong sample representation.

The final binary semantic differential scales confirm the findings of Figure 10.3, namely that respondents see a balance between visitors/tourists and local residents, thus not only recognizing St Godelina's Abbey for its local potential but also supporting a visitor economy. However, in this case, respondents lean towards a modest development with a smaller number of visitors. Finally, more potential is seen in a multi-destination/multi-activity offering, which might correlate with the need to accommodate both extra-local visitors as residents of Bruges and neighbouring municipalities, each having their own unique needs and interests.

Discussion

Analysing the results, it can be concluded that local residents want to preserve the historical narrative of St Godelina's Abbey in reconversion attempts and that novel activities should focus on providing spaces for rest, relaxation, contemplation, and small-scale exhibitions, training, and events – the latter particularly focusing on local artisans and artists. It is mostly envisioned as a community space where visitors are welcomed, although in modest numbers to avoid overcrowding and commercial exploitation. This is comparable to Lo Faro and Miceli's (2019) notion that the legacy of built religious heritage can be reallocated to new uses, pursuing social benefits for the community, protecting its strong bond with the collective memory, and addressing the objectives of social sustainability.

Even though the results of the participative inquiry are perhaps not surprising, with the future perspectives offered by individual participants largely coinciding with often-followed reconversion strategies, the case study can be seen as a success in terms of the participation rate, with a large majority of the 16,000 visitors willing to provide individual inputs. The high response rate for this type of study might indicate a general interest in the topic, and the site as a whole. This might have been further exacerbated by the fact that the site had been closed to the general public for centuries, and the short opening period during which visitors were allowed access in the summer of 2021 could have attracted highly motivated cultural visitors, as well as locals with a more vested interest in providing personal inputs for future development plans. It is also worth noting that the pilot project received much media attention, being featured on national television and radio, which will undoubtedly have contributed to visitor and survey participant numbers. Additionally, visitors to the site were invited to provide inputs via a questionnaire by staff members of VISITFLANDERS. Such face-to-face invitations are less likely to be ignored or rejected. While these elements combined to support a successful public survey in this particular example, at the same time they provide a caveat for extrapolation of the methodological operationalization. Not every reconversion project will be able to benefit from similar national (or regional) media attention, and budgets needed for face-to-face surveyors will often not be available. In such cases, the participatory approach to be followed might necessarily need to be of a more modest size.

Apart from the potential of large-scale participatory inquiries for generating novel reconversion ideas, as mentioned by Šegota et al. (2017), simply involving local society and keeping residents informed and included in tourism development projects can already help to garner support and a more positive perception of tourism. However, in order to avoid dissatisfaction, it is also essential that community inputs are at least somewhat incorporated in the final master plans. The extent of bottom-up implementation is at least partly linked to the level of participation that is afforded. Arnstein's (2019) ladder of citizen participation includes eight potential degrees, ranging from nonparticipation to tokenism, and finally citizen power:

(i) manipulation, (ii) therapy, (iii) informing, (iv) consultation, (v) placation, (vi) partnership, (vii) delegated power, and (viii) citizen control. In the case of St Godelina's Abbey, inhabitants and visitors were consulted, without handing over actual decision-making power. While this takes away potential initiative from local citizens, it can be more appropriate and realistic in cases where necessary investments are significant. For St Godelina's Abbey, the total investment needs are estimated at €14 million, including renovation and reconceptualization. Clearly, when such large financial needs are present, decision-making and management is better centralized, to work efficiently and retain accountability.

For the pilot site in Flanders, it meant that after the collection and subsequent analysis of individual responses, further strategies were explored by a panel of 40 experts. At this point in the process, one limitation became noticeable: of the 4,787 ideas that were analysed, only 238 were reflective of the "Travel to Tomorrow"/"Flourishing Destination" vision that is promoted by VISITFLANDERS. This is reflective of the two triangles of Figure 10.3 and the fact that a truly flourishing destination suggests a balance between people-place-activities and residents-entrepreneurs-visitors (i.e., respondents who locate themselves centrally in both triangles). Fully participatory approaches might thus create tensions between bottom-up ideas, and organizational strategic goals and objectives. This could, in turn, lead to a situation whereby either participatory inputs are largely ignored, or where an operationalization might not align with longer-term organizational strategies, particularly in cases where DMOs still hold primary responsibility over tourism development, whereas local participants might prefer non-tourism interventions. Furthermore, upon analysis it could be concluded that many scenarios remained rather superficial, indicating that most non-expert respondents might be better equipped to give insights into the general atmosphere, rather than suggesting development ideas.

The final architectural masterplan that was delivered through expert consultation contemporizes the historical narrative of the abbey while balancing local sustainability with economic profitability, conceptually overlapping with the main themes that were uncovered from the individual surveys. The final planned intervention includes eight different but related concepts: (i) a gastronomic concept with a cooking studio and demonstration kitchen, (ii) a creative open space for both planned and spontaneous cultural activities, (iii) an orchard with a participative collective artwork, (iv) a microbrewery and store for short supply chain produce, (v) a community garden with small-scale event space, (vi) temporary workspaces, (vii) entrepreneurial spaces for commercial organizations, and (viii) a housing project and meeting space for artists (VISITFLANDERS 2023). The varied, heterogeneous activity offer thus attempts to balance cultural visitor interests with open community spaces within a sustainable framework, fitting the "Travel to Tomorrow" vision of the DMO and also respecting the inputs delivered through the public inquiry.

Lastly, although the participatory approach followed in this pilot study, leading to the invitation of 16,000 participants and the eventual analysis of 4,784 responses, led to positive results in terms of local community integration and ideation, some critical remarks need to be made about the scope of the exercise and associated efforts needed. The research methodology set out to combine more qualitative inputs (i.e., open-text contributions), with a quantitative-like scope in responses. As briefly mentioned in the methodology section, questionnaires were respondent-completed on paper. These contributions were then digitalized by staff members of VISITFLANDERS for later analysis. It is worth noting that the initial estimate had been to collect 3,000 responses. However, partly due to the media attention being given to the initiative, the number of visitors – and as a result, questionnaire responses – was significantly higher than anticipated. This elevated the staff efforts needed for data input and analysis up to the point where it became unfeasible to process all responses. In the case of St Godelina's Abbey, the support from the DMO – both financially and in terms of staff contributions – has been significant, allowing for a study of this scope. However, in many other cases, it is worth considering whether the benefits of such an extensive public survey outweigh the costs, particularly when the intervention ideas resulting from public consultations are often relatively predictable and could be achieved by more targeted focus group discussions as well. The research set-up should therefore not be considered as a universal template and the level of public engagement needs to be balanced with the resources available and the scope of the interventions.

Conclusion

In contemporary times where tourism growth can potentially place unsustainable pressure on local natural, cultural, and social resources, DMOs have increasingly adapted their strategies from a historical growth-driven marketing approach towards a role of destination stewardship in which tourism is secondary to local well-being. In Flanders, Belgium, the DMO VISITFLANDERS has already had a longer history of balancing marketing with destination management. More specifically though, in 2017 the organization embarked on a deliberate co-creative process in order to update the role and responsibilities of the DMO, leading to a new vision: "Travel to Tomorrow".

While it has been a more common occurrence for DMOs to change their strategic visions and KPIs because of sometimes unsustainable tourism practices, operationalizing these concepts in tangible ways is still relatively new and requires both changes in marketing approach as well as changes in local development initiatives. This book chapter explored newly adopted strategies that include participatory approaches as a central component of tourism interventions. St Godelina's Abbey in Bruges, which has been a cultural-historic landmark for over 400 years, and which, partly due to increased secularization, needed revaluation in order to give it contemporary purpose, served as a pilot case. The participative approach took the form of a large-scale consultation of local inhabitants and visitors, analysing 4,784 future development scenarios over the course of just two months. The large bottom-up contribution was indicative of a significant interest in the future of the site, both among nearby residents and other visitors. Results indicated that people want to preserve a link between the historic past of the site, albeit in a non-religious fashion, protecting values linked with contemplation, sustainability, and connection. While commercialized, commodified touristic exploitation is not wanted, in general, there was support for open spaces that cater to visitors and locals alike. The public consultation provided valuable information to the DMO and an expert panel, to further elaborate a site masterplan.

Even though the public consultation allowed for the collection of a broad range of inputs, it is important to note that the research set-up was demanding in both staff involvement and financial coverage. The conversion and analysis of a large quantity of qualitative, textual inputs were labour-intensive and might therefore be prohibitive in the case of smaller reconversion projects. In such cases, more targeted consultations via focus groups might be preferable.

References

Arnstein, S. R. (2019). A ladder of citizen participation. *Journal of the American Planning Association*, 85(1), 24–34. https://doi.org/10.1080/01944363.2018.1559388.

Borzyszkowski, J. (2015). The past, present and future of destination management organizations (DMO) – The example of national tourism organizations (NTO). *Proceedings of the International Management Conference*, 9(1), 56–66. http://conferinta.management.ase.ro/archives/2015/pdf/6.pdf.

De Groote, P. (1999). *Panorama op toerisme. Handboek toerisme management in een internationaal perspectief*. Garant.

Dredge, D. (2016). Are DMOs on a path to redundancy? *Tourism Recreation Research*, 41(3), 348–353. https://doi.org/10.1080/02508281.2016.1195959.

Inventaris Onroerend Erfgoed (2023). *Sint-Godelieveabdij*. Retrieved 16 August 2023 from: https://id.erfgoed.net/erfgoedobjecten/82435.

Jainchill, J. (n.d.). From marketing to managing. Retrieved 22 August 2023 from: https://www.travelweekly.com/Travel-News/Government/destination-marketing-organizations-From-marketing-to-managing.

Lo Faro, A., & Miceli, A. (2019). Sustainable strategies for the adaptive reuse of religious heritage: A social opportunity. *Buildings*, 9(10), 211. https://doi.org/10.3390/buildings9100211.

Morgan, N. (2012). Time for "mindful" destination management and marketing. *Journal of Destination Marketing & Management*, 1(1&2), 8–9. https://doi.org/10.1016/j.jdmm.2012.07.003.

Morrison, A. (2013). *Marketing and managing tourism destinations*. Routledge.

Pechlaner, H., Zacher, D., Eckert, C., & Petersik, L. (2019). Joint responsibility and understanding of resilience from a DMO perspective – an analysis of different situations in Bavarian tourism destinations. *International Journal of Tourism Cities*, 5(2), 146–168. https://doi.org/10.1108/IJTC-12-2017-0093.

Šegota, T., Mihalič, T., & Kuščer, K. (2017). The impact of residents' informedness and involvement on their perceptions of tourism impacts: The case of Bled. *Journal of Destination Marketing & Management*, 6(3), 196–206. https://doi.org/10.1016/j.jdmm.2016.03.007.

VISITFLANDERS (2021). *Flora et Labora*. Toerisme Vlaanderen. Retrieved 21 August 2023 from: https://toerismevlaanderen.be/nl/over-ons/visie-strategie#floraetlaboradoc.

VISITFLANDERS (2023). *8 toekomstconcepten blazen Brugse Sint-Godelieveabdij nieuw leven in*. Retrieved 21 August 2023 from: https://www.samenherbestemmen.be/nl/nieuws/8-toekomstconcepten-blazen-brugse-sint-godelieveabdij-nieuw-leven.

Wang, Y. (2011). Destination marketing and management. In Y. Wang & A. Pizam (Eds.), *Destination marketing and management* (pp. 1–20). CABI.

Woodside, A. G., & Sakai, M. (2009). Analyzing performance audit reports of destination management organizations' actions and outcomes. *Journal of Travel & Tourism Marketing*, 26(3), 303–328. https://doi.org/10.1080/10548400902925379.

11 Murals as Creative Placemaking Interventions

The Case of Blind Walls Gallery Breda, The Netherlands

Marisa P. de Brito, Licia Calvi, Kristel Zegers, Josefien Boor, and Emma Braam

Introduction

The practice of wall painting and drawing can be traced back to prehistoric times, with pictographs in caves and ancient megalithic markings like those found in Egyptian tombs. The Romans utilized mosaics to convey power throughout their empire, while in modern history, government-funded artists such as Diego Rivera institutionalized muralism as a symbol of national identity. Today, luxury brands commission hand-painted murals for advertising (Mallon, 2023), and Banksy's street art is considered by two-thirds of the population to be a cultural symbol on par with the British royal family (Syer, 2018). Once strictly criminalized, signature graffiti is now acclaimed as street art (Abarca, 2016). Not surprisingly cities are increasingly adopting street art, such as murals, as a strategic tool to enhance their urban environments, often labelled as creative placemaking (Markusen & Gadwa, 2010).

Street art can serve as a visual storyteller of local culture, invoking a range of emotions and sentiments (Romero Jr., 2013). Literature examining the social impacts of art in public spaces highlights both positive and negative effects. On the one hand, advantages such as increased economic activity, urban revitalization, enhanced reputation, community pride, and overall development are often cited (Gratton & Preuss, 2013; Kaplanidou & Karadakis, 2010; Karadakis & Kaplanidou, 2012; Liu, 2016; Liu et al., 2014; Solberg & Preuss, 2007). On the other hand, street art can over time contribute to overcrowding and other inconveniences and is also blamed for accelerating gentrification processes, causing housing market prices to soar (Gratton & Preuss, 2013; Wijjedasa, 2020). Despite this, the presence of art in public spaces is generally viewed in a positive light (Zebracki et al., 2010), aligning with the belief that public art should have a positive social impact. Thus, many cities are using art as an intervention (Midgley, 2000) to elevate the urban space and researchers have pointed out the potential of urban culture for creative placemaking (Zuma & Rooijackers, 2020). However, there is limited research (Hall & Robertson, 2001), especially in the context of murals in inner cities.

In this chapter we focus on the city of Breda in the Netherlands as a case study. Breda boasts a collection of street art, featuring over 110 murals scattered throughout the city, collectively known as the Blind Walls Gallery (BWG) and referred to

DOI: 10.4324/9781003449027-13

as the museum of the street. This dynamic archive reshapes the cityscape while narrating the stories of its diverse neighbourhoods. Ultimately, it seeks to sensitize locals, tourists, and passersby, reshaping the way they relate to the city. Our exploration delves into the activities surrounding mural sites in two neighbourhoods in the city of Breda, and the significance attributed to them by residents.

Murals as Creative Placemaking Interventions

A place emerges from the social context wherein it is embedded – namely, the individuals who engage with it (Koopmans et al., 2017). This encompasses symbols, emotions, images, and ideas that forge a connection between these individuals and the place they inhabit. Consequently, a place becomes intertwined with social relationships and their subjective perceptions (Davenport & Anderson, 2005). As Tuan (1977, p. 4) puts it: "In essence, place consists of space infused with meanings and objectives through human experiences within this specific space." The concept of "sense of place" can be characterized as the individual's emotional and subjective response to a specific location (Tuan, 1974; Mulvaney et al., 2020), such as their neighbourhood.

Many urban authorities are using art purposefully to create change, along with what Midgley (2000) calls an intervention. Such processes of placemaking extend beyond the mere enhancement of spaces, aiming at transforming the significance and worth of a place (Cilliers et al., 2015). Public art is appealing as a tool to create inviting environments, conducive to interaction between passersby and residents (Guinard & Molina, 2018). By augmenting the visual allure of a city's grey alleys, it can attract tourists and expose specific areas to a broader audience (Grodach, 2010). It can also be a means to showcase and reinforce local identity, amplifying the symbolic significance of public spaces. This can be achieved through reflections on meaningful local episodes promoting the local identity (Zebracki et al., 2010).

Street art, such as murals, can be a vehicle for stacking meaning into a place since public spaces significantly shape the public's perception of their surroundings (Zebracki et al., 2010). Art in public space can result in a wide range of effects, from the way inner cities are viewed to the way they are used. Social impacts related to arts and culture are effects that extend beyond the direct experience, echoing into individuals' daily lives (Galloway, 2009). Among the impacts highlighted in the literature, one finds clusters around *urban regeneration*, with social, economic, and environmental aspects (Galdini, 2005), including improved perceived *image* and *safety*. Integrating arts and culture plays a pivotal role in enhancing social development at both individual and communal levels (Galdini, 2005; Garcia, 2004; Kay, 2000) within the urban regeneration process. Murals contribute significantly to the vibrancy of areas (Andron, 2018), often leading to reduced vandalism and crime rates. Artistic interventions can transform once unsafe spaces into secure environments (Sharp et al., 2005), signifying murals' positive impact on neighbourhood *safety*. Murals tend to elicit positive perceptions (Zebracki et al., 2010), potentially enhancing a neighbourhood's *image*. Such artistic creations have the potential to attract both inhabitants and visitors, fostering increased foot traffic that may lead to economic regeneration, particularly if businesses like shops and restaurants are nearby (Abarca, 2016; Grodach, 2010). Economic benefits are closely linked with tourism, and accessibility to murals proves pivotal in enticing visitors (Koster & Randall, 2005).

Conveying shared values, such as visual representations of important historical events or heritage, may strengthen community development (Hall, 2003). Thus, street art can showcase and reinforce local identity, amplifying the symbolic significance of public spaces, and nurturing social interactions (Miles, 1997). It also shapes the interaction with the environment. For instance, areas where artistic creations flourish tend to evoke a favourable impression from passersby, discouraging acts of vandalism (Andron, 2018). The infusion of art potentially adds value to the locale, enhancing its liveability and sociability, and in doing so enriching the sense of place for both residents and visitors (Doubleday, 2018).

The objectives of public art projects can vary (Sharp et al., 2005): they might aim at enhancing the environment's charm and may involve the public or residents in crafting such ambience. When the community actively contributes to these projects, a sense of shared responsibility for the outcomes is often cultivated (Sharp et al., 2005). Cultural planning strategies of local authorities are gradually shifting into creative placemaking: using arts and culture as a tool, side by side with the engagement of residents through a participatory approach (Zitcer, 2020; Markusen and Gadwa, 2010). The initiation of placemaking projects can thus be triggered in diverse ways. In a top-down approach, organizations like municipalities take the lead, while in a bottom-up approach, the impetus originates from the community itself, e.g., residents in a neighbourhood. Community engagement is assumed to be pivotal in placemaking, as alterations directly impact the inhabitants and users of these public areas. Ideally, placemaking is a collaborative endeavour aimed at shaping public spaces to enhance shared value (De Brito & Richards, 2017), while considering the social, cultural, and physical essence of a location (PPS, 2023)

There remains a noticeable void in research concerning murals as interventions in public space, and their impacts at the neighbourhood level. In this chapter, we use the case of Blind Walls Gallery in Breda to bridge this gap.

The Case: Blind Walls Gallery, Breda, the Netherlands

Breda – a city with 180,000 inhabitants in the south of the Netherlands – has a rich history, due to its importance in the Middle Ages as a residency of the Nassau dynasty and its function as a garrison town. Parts of this medieval history are still visible today in the historic city centre. Breda welcomes 1.7 million visitors on a yearly basis (Verlinden, 2019). The city does not suffer from overtourism, and the strategy of the city marketing office does not explicitly state attracting more tourists. Instead, the strategy is based on attracting and connecting visitors, businesses, and residents to create impact (Breda Marketing, n.d.).

In cultural terms, graphic design has a lengthy history in Breda with a renowned art school operating since 1947, a design museum (till 2017), many self-employed graphic designers, and a graphic design festival that has evolved into a continuous program organized by Graphic Matters. In the context of the graphic design festival, the BWG started. A triptych showing the city wall of medieval Breda was the inspiration for BWG to spray the first murals in 2014. Deliberately, the first murals were placed near spots where the city walls originally were, but now functioned as (dilapidated) parking lots. From the start, the murals were meant as temporary interventions, telling a story of one moment in the history of Breda that relates to the spot of the mural. These stories are selected by researching archives and talking to residents. The most inspiring stories are chosen and sent to selected (inter) national artists who are asked to reinterpret the story and create a sketch for the mural. Residents who live near the location, where the mural will be placed, are involved in the research of the stories, the selection of the artist, and the selection process of the sketches. This creates a relationship between the artists, their work, and the residents who see the mural every day. During the creation of the mural – which often takes several days – the artist stays in the BWG apartment, strengthening contacts with residents, the local art scene, and the BWG organization. Every mural is publicly unveiled with all stakeholders involved. The new mural is symbolically handed over to the residents who live closest to the artwork. In doing so, the mural becomes part of the public domain (BWG, 2022).

Every mural is thus the result of an intensive process balancing artistic quality and social interests. The stakeholder network consists of housing corporations/housing owners, project developers, residents, suppliers, funders, and artists. In 2021, the project evolved to an independent foundation with the following impact statement: "Blind Walls Gallery allows international artists to create meaningful murals in Breda. We share the stories behind the paintings and show how artists work. We encourage talent development and knowledge sharing" (BWG, 2022). It is BWG's belief that murals have a positive influence on the quality of life and social cohesion of the city. Furthermore, the BWG Foundation aims to ensure that murals are taken seriously as a form of art uniting artistic quality, social impact, and stimulating the professionalization of the art form as well as interdisciplinary collaboration (BWG, 2023).

Currently, the city has over 110 murals telling stories of Breda. These murals are scattered over Breda and connected by walking tours, biking tours, and publications. The BWG Foundation organizes paid guided tours (by foot, bike, step) for residents, visitors, students, and businesses in which the stories of the murals and the process of creating them are told. Yearly, 4,000 tickets are sold for guided tours. This is a small fraction of the total number of visitors which is estimated at (at least) 100,000 (BWG, 2022), since the murals are mostly visited in an unorganized way due to their location in public space.

To develop murals as an art form, knowledge about them is shared via mural expertise networks; BWG hosted and organized the first mural conference in 2022 in Breda. In the same year, it initiated a film festival on murals. Furthermore, the foundation advises other cities that include murals in public art strategies. Recently, BWG has been opened: a gallery in which artists who created one (or more) mural(s) exhibit and sell their work. BWG has its own merchandise including prints, caps, socks, books, and even bikes. A free city map containing the mural sites is the most wanted item by locals and visitors. These spread the word and have made the murals symbols of Breda.

Methodology

This research is part of a larger project focusing on tourism practices and the design of tourism interventions that cater to the well-being of those living around them. More specifically, in this chapter, we zoom into the studies of Braam (2022) and Boor (2022) which focused on the activities surrounding mural sites and the significance attributed to them by the residents. We revisit their work guided by the following question: What role do murals play in neighbourhoods, according to residents?

Two contrasting neighbourhoods were chosen in close cooperation with BWG (see Figure 11.1): (1) The Chasse neighbourhood, a rather residential area not far from the city centre, currently with two murals and no other touristic attractions; (2) Tuinzigt, a peripheral neighbourhood, west of the city centre, and where most of the murals in the city are located (seven in total), making it the most visited area of the city after the centre, among mural visitors

A qualitative approach was adopted by means of semi-structured interviews. Participants' selection criteria were as follows: (1) living close to the murals (or even in a building on which a mural is located), and (2) living in the neighbourhood

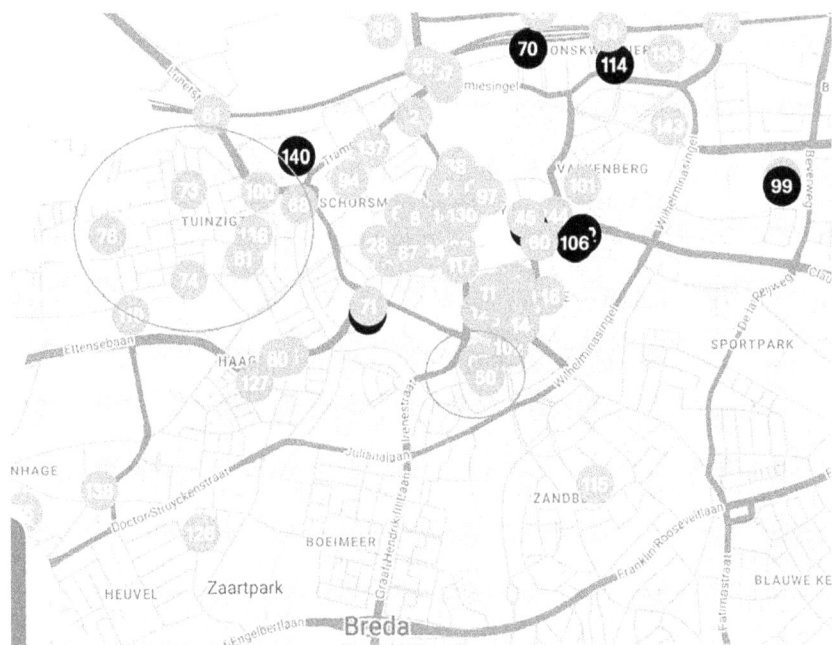

Figure 11.1 Murals in the city of Breda (the two selected neighbourhoods are circled)

for several years (to have witnessed any impact of the murals in the neighbourhood). Their age spanned from 20 to 75 and there were different degrees of involvement in the murals process, as will be explained per case.

Both studies applied saturation, with seven to ten respondents being interviewed per neighbourhood. An interview guide together with the list of topics was prepared in advance (Bryman, 2012), focusing primarily on practices (in the neighbourhood and around murals) and their meaning for the locals. More specifically, those topics included: (1) social interactions taking place in the neighbourhood (type of activities and by whom), (2) sense of place (views/feelings about the neighbourhood), (3) degree of involvement with the murals process, and (4) views/feelings towards the murals. Data was analysed through categorization and thematic analysis with the help of MaxQDA 2022. Quotes from the interviews will be used in the results and discussion sections, while using a pseudonym to preserve anonymity.

Results: Murals at Chasse Area

A subsection of the Chasse area has been chosen as the focus of the research. Within this section, two murals are located: one on a privately owned house created by Zenk One telling the story of mayor Godevaert Montens ("the lion") resisting the Spanish occupation in 1581. The second mural created by Collin van der Sluijs is located at the backside of the former Theatre Concordia which is located at the Concordia square. The painting was installed following the remodelling of the square, which now includes a small park and playground. The song thrush of the painting relates to the comic Abraham de Winter, who was formerly a popular artist in Breda. Apart from these two murals, there are no other touristic attractions in the area.

Chasse is a residential neighbourhood adjacent to the inner city, with some diversity. More than 55% of the inhabitants are autochthone or western migrants. Approximately 45% of the households own their house, 20% rent through social housing cooperation, and the remainder rent through the private sector (Statistieken buurt Chasse, 2023). The interviewed residents expressed their fondness of the neighbourhood due to its architecture, location, and the easiness of creating contacts. Within the studied area of Chasse, there is some degree of social cohesion: inhabitants are actively organizing street parties and monthly "cleaning up the neighbourhood" activities. Regarding the participation in setting up the mural, in our sample only a few interviewees have been actively involved. Respondents that were not present expressed – in hindsight – a wish to have been more involved from the start, as it would have been an opportunity to gain insight into choices, to obtain more information, and to guard the interest of the residents.

From the interviews, it could be derived that residents feel a sense of pride in their neighbourhood due to the murals. Murals are shown to others and respondents spontaneously showed pictures they'd taken of the murals to the interviewer saying: "I have seen him paint even. I have a photo of the creator as he puts his signature on it. That is kind of nice" (Ellen). Residents enjoy seeing visitors taking pictures of the murals.

Figure 11.2 Blind Walls Gallery, Zenk One

Source: Edwin Wiekens

Figure 11.3 Blind Walls Gallery, Collin van der Sluijs
Source: Rosa Meininger

The owners of the private house, where the Lion mural is, enthusiastically ex-plained that "we have been in newspapers and magazines everywhere", clearly showing a sense of pride and ownership towards the mural. For them the mural also serves as a point of identification: "If I then say with the mural of that lion, they already know where I live" (Viktor).

Residents believe that the two murals improved the image of the neighbour-hood. The colourful murals are perceived as making the neighbourhood more at-tractive. Not many residents were aware of the historical stories depicted in the murals. They consider the appeal of the murals as the foremost factor in revital-izing a once bleak wall. This said, other residents believe that more people are now aware of the history of their neighbourhood.

Respondents also point to safety: "I think it just brightens the place because often they are placed in places where you normally would not go, that you would normally even find a little scary when you walk past" (Bianca). Besides contribut-ing to feelings of safety, residents mentioned there is less misbehaviour in terms of graffiti or urinating in public.

The murals are believed to be intended as means of urban regeneration. Resi-dents express that the municipality and urban planners invest in the neighbourhood by providing space for art; they appreciate that bare and grey walls are transformed into pieces of public art. Furthermore, interviewees recognize that more people are attracted to the neighbourhood; they see more people using the park, and spend-ing time there chatting, eating ice cream, and playing. Also, foot traffic around the lion mural has increased because of the mural. The BWG tours bring more visitors to the Chasse area. At the time of writing, the residents had not experienced the increased number of visitors negatively.

Despite the rising numbers of visitors in the neighbourhood, there is no indication that this is resulting in economic benefits for the local bar. Residents explain that the bar owner had anticipated that the mural tours would encourage people to stop by for a drink at the café. However, it appears that the visitors use the bar primarily for the restroom facilities. Since having a mural in their own neighbourhood, residents are more open to discovering other murals in other neighbourhoods as well; this also changes the perception of inhabitants towards other parts of the city. Respondents mention that they are discovering parts of the city that they would not have visited previously, since there was no reason to before the murals were there.

Results: Murals in Tuinzigt

Tuinzigt is a rather large, relatively young (the largest age group is between 24 and 44 years old), and culturally diverse neighbourhood, with about 49% of its inhabitants having a migration background. About 40% of the houses are rented through social housing corporations, and about 25% are rented through the private sector, with the remainder being owned (Statistieken buurt Tuinzigt, 2023)

The neighbourhood is perceived by its residents, especially among those who have lived there for generations, as having a high internal social cohesion and solidarity: residents trust each other and feel safe there and engage in social activities like sitting together at their front door for a drink or a snack. On the days that the local soccer club plays, people put their flags out or walk together spontaneously to the stadium as mentioned in the following testimonial: "eventually you walk in a kind of joint parade, it's not arranged, it's unorganized and you walk together towards the stadium" (Timo).

A resident, who has lived their whole life there, describes how one can rely on their neighbours in case of need: "the people are still very close to each other. They help each other when there's a problem" (Kelly), while a newcomer describes living there as "just pleasant" (Joep). This is not that odd, as newcomers such as new students in town, may not have the same rituals, e.g., of supporting the local soccer team, and also expressed some initial prejudice towards the area. A long-time resident recognized that Tuinzigt is a stigmatized working-class neighbourhood with a negative image: "Well, that we as Tuinzigt do have a stigma, yes, many people say it is a slum, but they really do not know what it is like here. That is not the case at all" (Kelly).

A newcomer does partially let go of such an image: "If you live there, people are nice to you anyway" yet still adding, that despite the murals, "overall it is still a neighbourhood of row houses that have a bit of a drab appearance. That's how I would describe it at the moment still, unfortunately" (Joep).

Most of the interviewees appreciate the murals for their visual beauty, and for the value they represent. Of the seven murals present in Tuinzigt, the ones most

Figure 11.4 Blind Walls Gallery, Zenk One
*Source:*Edwin Wiekens

participants refer to are Zenk One and AlfAlfA, precisely because they clearly tell something about the history of Tuinzigt or of its residents. Zenk One is a mural dedicated to Ramon Dekkers, a world-famous kickboxer and child of Tuinzigt. This mural was very much wanted by the residents to celebrate their local hero. It came about as a grassroots initiative by many of the residents to replace an older memorial mural located at the place where Ramon died, which was in a dark tunnel and remote from his birthplace. People who worked hard to get the mural keep making sure it is maintained and not smudged. There is an overall sense of pride for it: they take pride not only in having managed to get it in place but also in the outcome. Due to Dekkers' fame worldwide, many sports organizations come to Tuinzigt just to see it as a tribute to Ramon. And not only them. Since the murals, the neighbourhood has become more colourful, busier, and livelier where spontaneous interactions occur between residents and tourists, the latter ones often asking for information on the murals. As accounted by the interviewees, the murals foster exchange among individuals of diverse backgrounds, as well as between home-owners and residents of social housing. The interactions also intensified with the newer residents, who were not so well connected before.

AlfAlfA, inspired by the many pigeon fanciers who lived around the area, including a former Dutch pigeon racing champion, has a less widespread appeal, being situated on a wall that no other buildings face, in a street with social houses

Figure 11.5 Blind Walls Gallery, AlfAlfA
Source: Edwin Wiekens

and houses from the 1930s. The community has not shown the same interest and involvement in bringing it to life as Zenk One. This mural came to life by the initiative of the owner of the house where it is painted. Nonetheless, the residents were also partly involved in the process through co-creative sessions to define what the mural should represent.

As a general remark, residents see an increase in people visiting their neighbourhood because the murals are gaining visibility in the city. They also notice an attitude change from the side of some of the residents who were initially not enthusiastic about the idea of having a mural on the wall of their building. At an earlier stage, they seemed to have considered murals akin to graffiti, associating them with vandalism and property damage. Now, it is a pretext to get to know better the history of the neighbourhood, as Jane attests: "I think it is cool that people … know a lot about each other. But little about the neighbourhood history … now they learn".

Discussion

The two neighbourhoods show some similarities in terms of their social dynamics despite being inherently relatively different due to their contrasting socio-economic fabric: Chasse has a higher rate of homeownership compared to Tuinzigt, where there is a significantly larger number of social housing rentals. Residents, especially the long-term ones, know each other rather well; there is social interaction and residents care for their neighbourhood. However, while in Chasse social interactions seems to be of a rather organized nature (e.g., neighbours organize the monthly clean-up of the neighbourhood, or organize festivities), in Tuinzigt the social interactions among residents seem to have more of a spontaneous character (e.g., coming together in front yards, parade to NAC football stadium).

There is a difference between the neighbourhoods in how the murals have intensified social interaction. In Chasse, residents especially discussed the murals with other residents when the artist was working on the mural. They do not report interacting with visitors, although one respondent mentioned she gave an extra piece of information about the mural to the BWG tour guide. All Chasse interviewees notice the visitors and appreciate the increased foot traffic, but this does not seem to lead to enhanced social interactions. On the contrary, in Tuinzigt, social interaction has intensified since the murals, both among residents, as well as between residents and visitors. The murals trigger conversations with the visitors, for example, when they seek directions to locate the murals, but also when they become intrigued about the mural's meaning or the residents' opinion of it. Tuinzigt residents therefore do experience a difference in the breadth and depth of the social interactions that take place since the murals have been installed. Now they entertain conversations, in the area where they live, with people they would not have had contact with before.

In both neighbourhoods, residents acknowledge that more visitors visit the area since the murals are there. Murals cause foot and bike traffic to the neighbourhood and inhabitants express that the number of visitors is currently good: they do not want to become an overcrowded tourist attraction.

It is interesting to note how residents talk about "their" murals, pointing towards the role murals can play, for example, in creating identity, enhancing image, strengthening a sense of belonging, and generating pride about, e.g., being featured in the press. Illustrative to this is the maintenance of the mural of the boxer by Tuinzigt residents: Ramon Dekkers is a celebrated child of Tuinzigt and their own hero (i.e., sense of identity and belonging), and some residents had intense involvement in the process of setting up the mural. Because of this, these residents feel very committed to it: they take care of its maintenance and make sure that the mural is not vandalized.

In this respect, we can acknowledge a difference in attitude towards the murals between the residents who have triggered the process of setting a mural up and those who have not. The former tend to feel prouder of the murals and show ownership of them, while the latter only point to improvements in the neighbourhood, such as having become more attractive, colourful, and lively. And this generally makes everyone happier.

The literature (Liu, 2016) distinguishes both positive and negative social impacts. In this research, no negative social impacts were reported by the locals. However, residents did voice a concern about the potential future increase in number of visitors: that when numbers continue to rise, this might cause inconvenience in their daily life. The city of Breda is not very touristic, and the increased number of visitors has not yet caused a nuisance in the eyes of inhabitants.

While there is undoubtedly a positive inclination towards the murals, this does not mean an immediate and comprehensive transformation in one's perspective of the neighbourhood. This said, residents do report pride, appreciation for improvement of the neighbourhood aesthetically, and a sense of safety. This is in line with the literature (Zebracki et al., 2010).

Conclusions

Given the research gap concerning murals as interventions in public space, and their impacts at the neighbourhood level, in this chapter we delved into the following question: What role do murals play in neighbourhoods, according to residents? This was done by means of a qualitative study in two neighbourhoods of the city of Breda, in the south of the Netherlands. Breda currently hosts more than 110 murals scattered around the city, as part of the Blind Walls Gallery, also known as the museum of the street. The interviews with residents focused on social interactions taking place in the neighbourhood, their degree of involvement with the murals process, and their views/feelings towards the neighbourhood in general, and the murals in particular.

Evidence indicates that the murals stimulated conversations among diverse groups of residents and fostered interactions between residents and passersby, ultimately promoting social engagement and cultivating a sense of pride and safety. Though the BWG murals have a transient character, whether some will transition from ephemeral to permanent will depend on the city redevelopment plans and the relationship between the locals and the murals as the murals potentially gain significance. To conclude that a handful of murals can entirely and instantaneously change the personal perception of a neighbourhood would be too far-fetched, especially in neighbourhoods with limited aesthetic appeal, or little greenery. Yet, in the case of the BWG interventions, murals go beyond the beautification of unappealing urban areas. The BWG murals are visual narrators of local identity, and at the same time they encourage social engagement and infuse fresh significance into locations, which can ultimately evolve into a distinct source of identity, as seems to be the case in the city of Breda.

We contend that murals possess the capacity to serve as placemakers by permeating a location with meaning. This significance is particularly pronounced when residents initiate or take a co-leadership role in the process. Public authorities should consider this when employing arts and culture for placemaking initiatives, emphasizing the importance of a participatory approach. It could even be argued that the role of authorities is more about facilitating and adopting a servant leadership approach in such endeavours. In addition, having murals in their own neighbourhood make residents more aware of the general presence of murals spread around the city, which may lead them to explore their surroundings more. Therefore, one could argue that these murals, while celebrating diversity, also foster unity among the various neighbourhoods in Breda.

It is crucial to acknowledge that the research occurred amidst the backdrop of increased public space usage due to the COVID-19 pandemic. Therefore, for enduring insights, longitudinal research is essential.

Acknowledgments

We would like to thank Blind Walls Gallery for the provided information and photos and their critical remarks on the draft version. Furthermore, we thank all the interviewees.

References

Abarca, J. (2016). From street art to murals: What have we lost? *SAUC-Street Art and Urban Creativity*, *2*(2), 60–67.

Andron, S. (2018). Selling streetness as experience: The role of street art tours in branding the creative city. *The Sociological Review*, *66*(5), 1036–1057.

Boor, J. (2022). The sense of place of neighborhoods filled with murals. A case study of Blind Walls in Tuinzigt Breda. Master's thesis, Breda University of Applied Sciences.

Braam, E. (2022). Social impacts in neighborhoods. Mapping the social impact generated by murals in a neighborhood. Master's thesis, Breda University of Applied Sciences.

Breda Marketing (n.d). Meerjarenstrategie Breda Marketing 2022–2026.

Bryman, A. (2012). *Social research methods*. 4th edition. New York: Oxford University Press.

BWG (2022). *Beleidsplan 2022*. Breda: Blind Walls Gallery

BWG (2023). *Beleidsplan 2023*. Breda: Blind Walls Gallery

Cilliers, E. J., Timmermans, W., van den Goorbergh, F., & Slijkhuis, J. S. A. (2015). The story behind the place: Creating urban spaces that enhance quality of life. *Applied Research in Quality of Life*, *10*(4), 589–598.

Davenport, M., & Anderson, D. (2005). Getting from sense of place to place-based management: An interpretive investigation of place meanings and perceptions of landscape change. *Society and Natural Resources*, *18*(7), 625–641.

De Brito, M. P., & Richards, G. W. (2017). Events and placemaking. *International Journal of Event and Festival Management*, *8*(1), 2–7.

Doubleday, K. (2018). Performance art and pedestrian experience: Creating a sense of place on the Third Street Promenade. *Geographical Bulletin*, *59*(1), 25–44.

Galdini, R. (2005). Urban regeneration process: The case of Genoa, an example of integrated urban development approach. *45th Congress of the European Regional Science Association: "Land Use and Water Management in a Sustainable Network Society"*, 23–27 August 2005, Amsterdam, The Netherlands, European Regional Science Association (ERSA), Louvain-la-Neuve.

Garcia, B. (2004). Urban regeneration, arts programming and major events: Glasgow 1990, Sydney 2000 and Barcelona 2004. *International Journal of Cultural Policy*, *10*(1), 103–118.

Gordon, K. (2013). Emotion and memory in nostalgia sport tourism: Examining the attraction to postmodern ballparks through an interdisciplinary lens. *Journal of Sport & Tourism*, *18*(3), 217–239.

Gratton, C., & Preuss, H. (2013). Maximizing Olympic impacts by building up legacies. In J. Mangan & M. Dyreson (Eds.), *Olympic legacies: Intended and unintended* (pp. 71–87). London: Routledge.

Grodach, C. (2010). Art spaces, public space, and the link to community development. *Development Journal*, *44*(4), 474–493.

Guinard, P., & Molina, G. (2018). Urban geography of arts: The co-production of arts and cities. *Cities*, *77*, 1–3.

Hall, T. (2003). *Art and urban change: Cultural geography in practice.* London: Edward Arnold.

Hall, T., & Robertson, I. (2001). Public art and urban regeneration: Advocacy, claims and critical debates. *Landscape Research, 26*(1), 5–26.

Kaplanidou, K. (2012). Legacy perceptions among host and non-host Olympic Games residents: A longitudinal study of the 2010 Vancouver Olympic Games. *European Sport Management Quarterly, 12*(3), 243–264.

Kaplanidou, K., & Karadakis, K. (2010). Understanding the legacies of a host Olympic city: The case of the 2010 Vancouver Olympic Games. *Sport Marketing Quarterly, 19,* 110–117.

Karadakis, K., & Kaplanidou, K. (2012). Legacy perceptions among host and non-host Olympic Games residents: A longitudinal study of the 2010 Vancouver Olympic Games. *European Sport Management Quarterly, 12*(3), 243–264.

Kay, A. (2000). Art and community development: The role the arts have in regenerating communities. *Community Development Journal, 35*(4), 414–424.

Koopmans, M., Keech, D., Sovova, L., & Reed, M. (2017). Urban agriculture and placemaking: Narratives about place and space in Ghent, Brno and Bristol. *Moravian Geographical Reports, 25*(3), 154–165.

Koster, R., & Randall, J. E. (2005). Indicators of community economic development through mural-based tourism. *Canadian Geographer/Le Géographe Canadien, 49*(1), 42–60.

Liu, D. (2016). Social impact of major sports events perceived by host community. *International Journal of Sports Marketing and Sponsorship, 17*(1), 78–91.

Liu, D., Broom, D., & Wilson, R. (2014). Legacy of the Beijing Olympic Games: A non-host city perspective. *European Sport Management Quarterly, 14*(5), 485–502.

Mallon, J. (2023, 23 March). Luxury brands choose hand-painted murals for advertising, Fashion United. *Fashion United.* https://fashionunited.com/news/fashion/luxury-brands-choose-hand-painted-murals-for-advertising/2023032052936.

Markusen, A., & Gadwa, A. (2010). *Creative placemaking.* White Paper for the National Endowment for the Arts. http://arts.gov/pub/pubDesign.php.

Midgley, G. (2000). *Systemic intervention: Philosophy, methodology, and practice.* New York: Kluwer Academic/Plenum Publishers.

Miles, M. (1997). *Art, space and the city: Public art and urban futures.* London: Routledge.

Mulvaney, K. H., Merrill, N., & Mazzotta, M. (2020). *Sense of place and water quality: Applying sense of place metrics to better understand community impacts of changes in water quality.* London: IntechOpen.

PPS (2023). What is placemaking? *Project for Public Spaces.* https://www.pps.org/category/placemaking.

Romero Jr., A. (2013, 24 May). Murals take on new form and purposes. *The Intelligencer,* p. 3.

Schensul, J. J., & Trickett, E. (2009). Introduction to multi-level community based culturally situated interventions. *American Journal of Community Psychology, 43*(3–4), 232–240.

Sharp, J., Pollock, V., & Paddison, R. (2005). Just art for a just city: Public art and social inclusion in urban regeneration. *Urban Studies, 42*(5–6), 1001–1023.

Solberg, H. A., & Preuss, H. (2007). Major sport events and long-term tourism impacts. *Journal of Sport Management, 21*(2), 213–234.

Syer, J. (2018, 5 October). Banksy: A national treasure? *MyArtBroker.* https://www.myartbroker.com/.

Statistieken buurt Chasse (2023, 1 October). Buurt Chassé (gemeente Breda) in cijfers en grafieken (bijgewerkt 2023!), *AlleCijfers.* https://allecijfers.nl/buurt/chasse-breda/.

Statistieken buurt Tuinzigt (2023, 1 October). Buurt Tuinzigt (gemeente Breda) in cijfers en grafieken (bijgewerkt 2023!), *AlleCijfers.* https://allecijfers.nl/buurt/tuinzigt-breda.

Tuan, Y.-F. (1974). *Topophilia: A study of environmental attitudes, perceptions and values.* Englewood Cliffs: Prentice Hall.

Tuan, Y.-F. (1977). *Space and place: The perspective of experience.* Minneapolis: University of Minnesota.

Verlinden, P. (2019, 14 November). Niet koste wat het kost meer bezoekers naar Breda halen. *BndeStem.* https://www.bndestem.nl/breda/niet-koste-wat-kost-meer-bezoekers-naar-breda-halen~a2443d71d/.

Wijjedasa, I. (2020, 9 October). Beauty and pain: Street art's contribution to gentrification. *The Terrace.* https://theterracebc.com/2020/10/09/beauty-and-pain-street-arts-contribution-to-gentrification/.

Zebracki, M., Van der Vaart, R., & Van Aalst, I. (2010). Deconstructing public artopia: Situating public-art claims within practice. *Geoforum, 41*(5), 786–795.

Zitcer, A. (2020). Making up creative placemaking. *Journal of Planning Education and Research, 40*(3), 278–288.

Zuma, B., and Rooijackers, M. (2020). Uncovering the potential of urban culture for creative placemaking. *Journal of Tourism Futures, 6*(3), 233–237.

Part III

Critical Analysis of Interventions as a Steppingstone for Future Success

12 Economic Impact of Niche-based Tourism

Case Study of Off-Highway Vehicle (OHV) Recreation

Eunhye Grace Kim and Deepak Chhabra

Introduction

Off-highway vehicle (OHV) recreation is becoming an increasingly popular recreational activity which provides both social and economic benefits (Hughes & Paveglio 2019). OHV is any vehicle intended to be ridden off-highway, including all-terrain vehicles (ATVs), utility task vehicles (UTVs), side-by-sides, recreational off-highway vehicles (ROVs), dune buggies, sand rails, motorcycles or dirt bikes, and snowmobiles (Kil et al. 2012). OHV activity is one of the fastest growing segments of recreation in the U.S. (Hughes et al. 2014). The state of Arizona has plentiful natural resources suitable for OHV recreation. OHV trails are multiple paths available for OHVs used for recreational purposes. OHV trails in the U.S. are managed by federal, state, local, and/or tribal agencies as well as private organizations (Arizona State Parks 2018).

This study is part of a larger project titled "Economic impact of off highway recreation in the state of Arizona". The project aimed to provide marketing information about the OHV visitors in Arizona and present a comprehensive analysis of the economic impact.

The study's method is based on analysis of online surveys collected in 2017. The data was collected to gather information from visitors to understand travel and recreation motivations, travel and visit behavior, activity preferences, spending, and satisfaction levels so that Arizona State Parks can improve future service (Arizona State Parks 2018). The economic impact analysis uses spending and travel behavior data from that study to measure how spending by an OHV visitor circulates through local gateway economies, generating business sales and supporting local jobs and resident income.

OHV recreation has developed significantly in terms of number of users. Several studies expected that the popularity of OHV recreation on U.S. public lands will continue to grow (Bowker et al. 2012; Cordell et al. 2008; Kil et al. 2012). These projected increases will likely require federal, state, and local governments to consider additional management standards (Bosworth 2004; Monz et al. 2010). OHV recreation presents an opportunity to enhance local economies near popular OHV destinations through increasing tourism activities (Hughes et al. 2014; Hughes & Paveglio 2019). Hughes et al. (2014) indicate that OHV recreation

DOI: 10.4324/9781003449027-15

generates a potentially lucrative market for rural communities desiring to diversify their economy.

Several studies report that OHV recreation produces substantial economic benefits to the communities and to the state in which the recreation activity takes place (Cordell et al. 2008; Legg et al. 2006; Otto 2008; Silberman & Andereck 2006). OHV users spend money on food, lodging, and other travel-related expenditures. The recreational equipment they use stimulates economic activity and employment. They significantly contribute to the quality of life for local residents. The spending associated with the use of OHV trails and recreation is also important to the local economy as it generates jobs and income (Arizona State Parks 2018).

The research project was conducted to measure the economic impact of OHV recreation by retained and out-of-state visitors in Arizona. Retained visitors are defined as locals who listed an alternative park outside their study area they would have been interested in visiting if the OHV recreation was not offered in Arizona. This implies that the OHV trails retain outside spending and should be factored into the economic impact (Arizona State Parks 2018). To identify the local retained visitors, the survey asked a question to those who mentioned that they are residents of Arizona: If the OHV trails were not available, would you have traveled outside of Arizona for OHV recreation? Those who answered "Yes" to this question were categorized as retained visitors. The particular study for this chapter focuses on the most popular OHV vehicle, UTV/side-by-side, as identified by the survey results, while the primary reason for the OHV trip is trail riding using a UTV/side-by-side.

Economic Effects Analyses

To determine economic benefits of OHV recreation, a more comprehensive economic impact study is required which makes use of an input/output modeling technique focused on social accounting matrices, multipliers, and trade flows (Stynes 2010). First, social accounting matrices (SAMs) consider real dollars of all business transactions happening annually in the economy as reported by businesses and governmental agencies. SAMs also factor in "non-market" transactions such as taxes and unemployment benefits. This type of analysis provides an in-depth look at the economic impact of visitor expenditures on the local economy and can identify different types of industries and households that benefit the most from the economic impact analysis (IMPLAN 2014). Secondly, multipliers help to show direct, indirect, and induced impacts because of the visitor spending based on 536 different industries. Multipliers improve the accuracy of economic impact studies by calculating how the initial expenditures injected in the region can spur additional/indirect purchases of goods and services to meet demand for tourism products (Crompton 2010; Greenwood & Vick 2008; Gunthar et al. 2011). Direct effects are determined by the initial visitor expenditures injected into the economy. The indirect effects relate to additional spending incurred by tourism businesses in the region to produce/supply the tourism products. Examples include expenditures incurred by restaurants on groceries, services, other supplies, and labor. Finally, the induced effect measures the money that is re-spent in the study area as a result of employee spending (drawn from income earned as a result of direct and indirect effects). Lastly, the trade flows method helps to capture regional purchase coefficients and can perform a multi-regional analysis to show an organization's product can generate additional effects in the surrounding areas (IMPLAN 2014).

This economic impact study examines the direct, indirect, and induced economic effects of the OHV visitor expenditures in the gateway region of each selected trail and for the whole state. Visitors/tourists in this study are broadly categorized as: retained (locals), in-state non-locals or domestic (from outside the gateway region but residing in Arizona), and out-of-state. Spending (retained) is considered for those residents who would have gone elsewhere if the OHV trail they were using had been absent. IMPLAN software is used, which to date is the most used software to determine economic impacts of recreation and tourism in the United States.

Information is also elicited on the following: day tripper versus overnight visitors, length of stay, place of stay, place of residence (zip code), and retained (spending of local residents who would have traveled outside the state if OHV trails had been absent) or displaced (substitute spending that might have been incurred on items other than those associated with the OHV recreation) expenditures (Arizona State Parks 2018).

Methodology

Types of Economic Effects Measured

The economic effects of visitor spending to local economies are estimated by multiplying visitor spending by regional economic multipliers. Five types of economic impacts are used for this study: employment, labor income, value added, output, and indirect business taxes and tax impacts. Employment refers to annual average jobs. This includes self-employed and wage/salary employees, and all full-time, part-time, and seasonal jobs, based on a count of full-time/part-time average over 12 months. Labor income is composed of two components. These are employee compensation and proprietor income. Employee compensation is total income to the labor factor of production. From the point of view of a business, employee compensation is the total cost of labor including wages and salaries, other labor-related income like health and retirement benefits, and both employee and employer contributions to social insurance. Proprietor income is the total income to a sole proprietor or self-employed "employee". Value added is the combination of labor income, other property type income, and indirect business taxes. Other property type income includes corporate profits, interest income, and rental payments. Value added accounts for all non-commodity payments associated with an industry's production. Output can be described as the total value of production. Indirect business taxes are taxes collected by businesses on behalf of the government. These include sales tax, excise tax, property tax, fees, fines, and licenses. Tax impacts are categorized as federal and state/local. The tax impact report provides information on tax collection by state/local and federal governments. The software does not break state taxes from county taxes in a region but if the impact region is local, then state/local tax implies local tax contributions and jobs.

Robusticity of Web-Based Surveys

Web-based surveys have been noted as one of the most significant advances in survey methodology despite some limitations associated with computer anxiety, interface issues, and different screen formats. Advantages include: (1) low cost, fast response time, and flexibility; (2) interactivity without interviewer bias; (3) target sample selection; (4) quick and easy access; (5) desensitizing sensitive subjects, and (6) fewer processing errors. Online surveys can be conducted very quickly and efficiently. The respondents can conveniently complete the survey at their own pace. Also, the interactivity nature of the Internet reduces possible bias due to the absence of the interviewers. It is easier to accurately select respondents, e.g., with a pop-up invitation window when certain responses are recorded by the system. Also, respondents can be conveniently directed to questions of relevance. One key advantage is the ability to effectively reach respondents across different geographic locations. Additionally, being able to complete an online survey with privacy and convenience can minimize embarrassment and sensitivity towards certain controversial issues or personal topics. Online surveys can also be used to collect information related to unpopular beliefs or attitudes. In addition, online surveys provide stronger anonymity when compared to other survey techniques. Lastly, responses are recorded online simultaneously. With careful design, typical data input and processing errors can be avoided (Arizona State Parks 2018).

A comprehensive online survey was designed to accomplish the study objectives. During the development process, existing studies were examined to determine different spending items associated with OHV recreation. The survey link was hosted at the Arizona State Parks' website. Additionally, visitors at geographically dispersed trails were randomly surveyed at selected popular trail sites. The visitor data was used to segment visitors by type of trip. OHV visitors were split into the following three distinct visitor segments in order to help explain differences in spending across different user groups: (1) local visitors; (2) local retained visitors – who would have traveled outside of Arizona for OHV recreation if the OHV trails were not available in Arizona (day trippers and overnight visitors); (3) out-of-state, non-local visitors from outside of Arizona (day trippers and overnight visitors).

Sampling Design

Total OHV recreationists (out-of-state and retained locals) for the financial year 2017 was calculated based on a weighted percentage of OHV users, reported from a study conducted by Budruk et al. (2014). The report indicated that 12.6% of the local population of Arizona are OHV users. Efforts are made to obtain the following estimates: total expenditures for day and overnight trips; annual expenditures to purchase and maintain vehicles; other expenditures supporting the recreational use of vehicles; economic contributions to the state of Arizona; total expenditures made by OHV recreationists; and number of day trips and overnight trips by residents and non-residents by OHV type (such as ATVs, dirt or dual-purpose bikes, snowmobiles, 4WDs, side-by-sides, dune buggies/sand rails). Expenditure categories include both trip and annual expenditures. Trip-related expenditure items include gasoline/oil for motorized recreational vehicles and tow vehicles; restaurants/lounge purchases; food and beverages purchased at grocery and/or convenience stores; user fees and donations; guides and tour packages; lodging (overnight trips only); repairs, souvenirs; gifts and entertainment; and other trip-related expenditures. Itemized vehicle expenditures include: maintenance, repairs, storage, and miscellaneous items in Arizona needed by resident and non-resident households that participate in motorized recreation included repairs and parts; vehicle storage; motorized recreational accessories (covers, saddle or tank bags, ski skins, studs, carbides, mirrors, etc.); clothing (suits, pants, gloves, etc.); safety equipment (helmets, tools, first aid, etc.); annual insurance payment, registration, or permit fees; club memberships; magazine subscriptions; and vehicle purchase. Table 12.1 summarizes the types of vehicles and expenditure items asked for in this study (in regard to trip-related expenditure items, respondents were asked to answer based on the most recent trail visited, and for itemized vehicle expenditures, they were asked to answer based on the annual vehicle-related expenditures).

IMPLAN modeling software was used to calculate the economic impact to determine direct, indirect, and induced effects of visitor spending in terms of output, value-added, and labor income. IMPLAN software is the most used software to determine the economic impact of recreation and tourism in the United States. Several studies have used the IMPLAN input/output model to capture the direct, indirect, and induced impacts of OHV recreation to state economies (Kaliszewski 2011; Pardue & Shand 2014; Taylor et al. 2013). The currently available IMPLAN Version

Table 12.1 Types of vehicles and expenditure items used in questionaries

OHV type	Trip-Related Expenditure Items (Ancillary Expenditure)	Itemized Vehicle Expenditures (Operating Expenditure)
ATV	Gasoline/oil for OHV and tow vehicles	Motorized recreational accessories (trailers, covers, add-ons, ski skins, etc.)
Dirt bike	Restaurant/lounge purchases	Annual insurance payment
Dual sport bike	Food and beverage purchased at grocery and/or convenience stores as part of OHV recreation trip	Licenses, registration or permit fees
Snowmobiles	Guides and tour packages	Sticker fund
4WD	Lodging (overnight trips only) as per part of OHV activities	Club memberships
Side-by-side	Repairs/maintenance	Purchase of OHVs
SUV/Jeep	OHV rentals	Fuel, oil, and lubricants (OHV)
Dune buggy/Sand rail	Sporting goods	Other expenses
	Souvenirs and gifts	
	Amusement/entertainment	
	Clothing (helmets, boots, gloves, etc.)	
	Other OHV trip-related expense	

3 modeling system uses 536 distinct sectors and can offer a summary of economic impact in the county, in terms of changes in jobs, household income, tax impacts, and gross regional product, as new expenditures are injected into the economy. It is used to measure the impact of visitor expenditures on local economies in terms of output, value added, labor income, and federal and state/local taxes. The IMPLAN software system uses the input/output modeling technique to understand how a local economy functions and the economic benefits of local parks and recreational activities/facilities (Arizona State Parks 2018). Additionally, it makes use of an input/output modeling technique focused on social accounting matrices, multipliers, and trade flows (Greenwood & Vick 2008; IMPLAN 2014). Operating expenses are a breakdown of expenses involved with the operation of OHV vehicles, and the survey requested respondents to share their annual total OHV operating expenses in order to view operating expenses under the itemized vehicle expenses. For the purpose of economic impact, the breakdown of visitors based on each OHV type is computed from the number of visitors who offered vehicle spending data.

Findings

A total of 3,550 online surveys were collected in the year 2017 with 1,654 completed surveys. Total number of onsite surveys are 142. Based on the weighted percentage of OHV users, reported from a study conducted in 2014 (Budruk et al. 2014), 12.6% of the local residents of Arizona are reported to be OHV users. According to the online survey results, 90% of the OHV visitors are from Arizona and 10% are from out of state. The total population of Arizona, according to the 2017 census data, is 7,016,270. Based on this population, a total of 792,109 local OHV users was calculated. Next, a total of 890,336.8 OHV users are ascertained by adding 10% of out-of-state OHV visitors. Retained OHV visitors total 554,413 of which 372,566 are day trippers and 181,848 are overnight visitors. A total of 98,228 out-of-state visitors were calculated of which 39,488 are day trippers and 58,740 are overnight visitors.

Table 12.2 presents the total direct expenditures incurred by OHV visitors in 2017. As the table illustrates, OHV visitors in the state of Arizona generated a total annual direct spending of approximately $1.86 billion in 2017. Contributions from annual ancillary items are approximately $793.41 million. Annual vehicle expenses across all OHV vehicles are $1.07 billion. Tables 12.3–12.5 show detailed spending information of ancillary items and operating expenses by retained local visitors and out-of-state visitors.

Table 12.2 Aggregate visitor expenditures for all OHV users

Item	Ancillary	OHV Operating Expenses	Total Expenditures
Retained	727,897,419.02	888,684,893.68	1,616,582,312.70
Out of State	65,512,121.01	177,606,602.80	243,118,723.81
Total	793,409,540.03	1,066,291,496.48	1,859,701,036.51

Table 12.3 Annual total spending of retained local OHV visitors on ancillary items

Item	Day Trippers	Overnight Visitors	Total Spending ($)	Percent
Gasoline/oil for OHV and tow vehicles	123,844,664.06	15,178,852.56	139,023,516.62	19.10
Restaurant/lounge purchases	70,958,920.36	13,154,884.32	84,113,804.68	11.56
Food and beverage purchased at grocery and/or convenience stores	54,409,538.64	23,647,513.92	78,057,052.56	10.72
Guides and tour packages	4,698,057.26	1,827,572.40	6,525,629.66	0.90
Repairs/maintenance	116,985,724.00	21,676,281.60	138,662,005.60	19.05
OHV rentals	1,758,511.52	589,187.52	2,347,699.04	0.32
Sporting goods	36,153,804.64	10,083,471.60	46,237,276.24	6.35
Souvenirs and gifts	11,162,077.36	2,431,307.76	13,593,385.12	1.87
Amusement/entertainment	20,070,130.42	2,745,904.80	22,816,035.22	3.13
Clothing (helmets, boots, gloves etc.)	50,240,525.10	6,884,765.28	57,125,290.38	7.85
Other	8,829,814.20	22,552,788.96	31,382,603.16	4.31
Lodging – Camping/RV	47,315,882.00	10,001,640.00	57,317,522.00	7.87
Lodging – Hotel/motel	44,301,823.06	6,393,775.68	50,695,598.74	6.96
Total	590,729,472.62	137,167,946.40	727,897,419.02	100.00

Table 12.4 Annual total spending by out-of-state visitors on ancillary sectors

Item	Day Trippers	Overnight Visitors	Total Spending ($)	Percent
Gasoline/oil for OHV and tow vehicles	8,725,268.48	4,481,862.00	13,207,026.00	20.16
Restaurant/lounge purchases	5,873,050.24	3,331,145.40	9,203,961.47	14.05
Food and beverage purchased at grocery and/or convenience stores	4,343,285.12	3,701,207.40	8,044,494.60	12.28
Guides and tour packages	211,260.80	67,551.00	278,828.20	0.43
Repairs/maintenance	5,741,160.32	3,544,371.60	9,285,612.44	14.17
OHV rentals	419,757.44	128,053.20	547,775.60	0.84
Sporting goods	1,115,141.12	890,498.40	2,005,621.34	3.06
Souvenirs and gifts	958,768.64	844,093.80	1,803,077.89	2.75
Amusement/entertainment	1,357,202.56	1,705,809.60	3,585,767.22	5.47
Clothing (helmets, boots, gloves etc.)	2,290,698.88	1,515,492.00	3,806,467.00	5.81
Other	3,080,853.76	1,802,143.20	4,883,013.48	7.45
Lodging – Camping/RV	2,292,673.28	2,629,789.80	4,922,584.00	7.51
Lodging – Hotel/motel	1,535,688.32	2,402,466.00	3,937,891.77	6.01
Total	37,944,808.96	27,044,483.40	65,512,121.01	100

Table 12.5 Annual total vehicle-related spending

Item	Retained AZ Users		Out-of-State OHV Visitors	
	Total Spending ($)	Percentage	Total Spending ($)	Percentage
Motorized recreational accessories (trailers, covers, add-ons, etc.)	170,694,308.13	19.27	29,448,953.9	16.6
Annual insurance payment	51,658,232.06	5.83	7,454,786.8	4.2
Licenses, registration/ permit	17,396,118.66	1.96	3,099,375.5	1.7
Sticker fund	2,055,960.47	0.23	287,823.5	0.2
Club memberships	2,318,170.33	0.26	468,141.6	0.3
Purchase of OHVs	540,999,088.51	61.08	122,723,431.8	69.1
Fuel, oil, and lubricants (OHV)	83,638,294.60	9.44	12,609,539.8	7.1
Other expenses	16,904,481.07	1.91	1,514,549.8	0.9
Total	885,664,653.83	100.00	177,606,602.70	100.0

Table 12.6 Economic impact of OHV expenditures ($)

Impact Type	Employment	Labor Income	Value Added	Output
Direct Effect	5,572.74	341,420,128	587,895,749	888,684,878
Indirect Effect	2,342.84	119,147,907	196,786,965	353,697,933
Induced Effect	3,265.25	145,762,361	261,206,508	457,294,654
Total Effect	11,180.83	606,330,396	1,045,889,222	1,699,677,465

Table 12.6 presents the economic impact contributions of OHV expenditures. As Table 12.6 shows, OHV spending generated $1.70 billion in output, $1.04 billion in value added, and 11,180 jobs.

Tables 12.7 and 12.8 present economic impact contributions of total ancillary spending by OHV visitors in terms of jobs, labor income, value added, output, and state/local and federal tax contributions.

When asked the reason for OHV trip, trail riding UTV/side-by-side was found to be most popular reason (28.7%). With the marketing profiles, the most popular OHV vehicles are UTV/side-by-side (42.9%).

Operating expenses allow for a breakdown of expenses associated with the operation of OHV vehicles by asking annual expenditures on OHV vehicle-related items. For the purpose of economic impact, breakdown of visitors based on each OHV type is computed from the number of visitors who offered vehicle spending data. According to the retained local visitors' data from the itemized breakdown of total spending, the users of UTV/side-by-side indicate the highest vehicle-related spending (total $413,481,699). Also, UTV/side-by-side users (median party size = 4) indicate the highest average expenditure among the vehicles (total average: $2,785.42) which is 29.39%. Based on OHV operating expenses by out-of-state visitors, UTV users (median party size for UTV/side-by-side = 5) show the second

Table 12.7 Economic impact breakdown ($)

	Employment	*Labor Income*	*Value Added*	*Output*
Direct	6,901.70	197,543,169	306,591,804	502,580,471
Indirect	1,183.60	56,091,950	102,954,734	182,510,630
Induced	1,811.00	80,354,422	143,979,375	252,275,611
Total	9,896.30	333,989,541	553,525,913	937,366,712

Table 12.8 State/local and federal taxes 2017 ($)

Tax Category	Employee Compensation	Proprietor Income	Tax on Production	Households	Corporations	Total
Total State & Local Tax	0.00	0.00	66,199,520	6,065,089	1,014,070	73,278,679
Total Federal Tax	37,575,928	1,203,013	8,388,784	22,548,532	9,568,805	79,285,062
Total	37,575,928	1,203,013	74,588,304	28,613,621	10,582,875	152,563,741

Table 12.9 Average spending on OHV vehicle operating expenses ($)

Vehicle	Retained		Out of State	
	Average Spending	*Percent*	*Average Spending*	*Percent*
Dune Buggy	343.81	3.63%	484.39	4.8
SUV/Jeep	1,273.08	13.43%	1,090.31	10.9
UTV/side-by-side	2,785.42	29.39%	2,620.24	26.1
4WD	1,509.33	15.93%	1,914.67	19.1
Dual Purpose Sport Bike	995.33	10.5%	385.54	3.8
Dirt Bike	1,408.05	14.86%	2,712.34	27.0
ATV	784.70	8.28%	839.73	8.4
Snowmobile	376.36	3.97%	0	0
Total	9,476.08	100.0%	10,047.22	100

highest total operating expenses. Table 12.9 presents average spending information for each person.

Discussion and Limitations

Outdoor recreation is increasingly of strategic interest to rural areas seeking to encourage economic development through increased outdoor recreation-linked tourism (EPA 2019; White et al. 2016). Outdoor recreational activity generates spending and economic activity in local communities, and its amenities support the quality of life and health of individuals, communities, and local economies (Duval et al. 2020). OHV recreation has experienced remarkable growth in popularity in the U.S. for the past few decades (Cordell et al. 2008; Cordell et al. 2005; Hughes & Paveglio 2019; Jakus et al. 2010; Kil et al. 2012; Magnini et al. 2022).

OHV users have also been identified as a recreational group of high spending on travel-related expenses (Thomas et al. 2008; Holmes & Englin 2010).

As this study confirms, it is beyond a doubt that OHV recreation continues to grow and provide economic benefits for the nation (UTV Action 2022), and OHV recreation is a proven economic stimulus to the tourism market (UTV Action 2019). In 2022, the U.S. Department of Commerce's Bureau of Economic Analysis (BEA) released the report "Outdoor recreation satellite account, U.S. and states, 2021." According to the BEA, the data shows that outdoor recreation generates $862 billion in economic output and 4.5 million jobs (3% of employment), and inflation-adjusted gross domestic product for the outdoor recreation economy increased 18.9% last year, easily topping the 5.9% increase for the overall U.S. economy (Bureau of Economic Analysis 2022). The National Off-Highway Vehicle Conservation Council (NOHVCC) also reports that the economic impact of OHV recreation is $68,000,000,000 (UTV Action 2019).

UTVs are growing in popularity, and manufacturers are investing on television and print media ads to attract new customers (UTV Action 2022). The BEA listed ATV/UTV riding (which includes side-by-sides) as the nation's fourth largest "conventional outdoor recreation" activity for 2020–2021, with $8.1 billion in current-dollar value added for the economy – up from $7.3 billion the previous year (Bureau of Economic Analysis 2022). The data from the BEA confirms exactly how the outdoor recreation industry has created a strong and resilient base for the country's economy (UTV Action 2022). According to NOHVCC, UTVs account for 44% of OHV sales in the South, 27% in the Midwest, 21% in the West, and 8% in the Northeast (UTV Action 2019).

As a result of COVID-19, Arizona State Parks had to limit some of the OHV services including large-scale events and tours. Some facilities also had to be closed for guest safety. However, OHV use has exploded in popularity during the pandemic. Arizona State Parks have seen increased use across the entire state and several of the OHV retailers have seen increased sales. It can be concluded that OHV recreation allows the public to recreate themselves while social distancing in the midst of a global pandemic.

Despite this growth, trail access and development for OHVs has grown at a slower rate than demand (Center for Business and Economic Research 2006). These conditions make OHV development a potentially lucrative market for rural communities, particularly those close to public lands that are suitable for OHV recreation. Given the continued increase in OHV use, this study confirms the potential economic impact associated with OHV tourism. As Hughes et al. (2014) indicate, management strategies need to be carefully considered to produce attractive and sustainable OHV trail systems.

Like all studies, this study is also subject to logistical and methodological limitations. The self-reported responses of online surveys are voluntary and run the risk of excluding certain populations who do not have access to the Internet and might be in the low-income category. Although the representativeness of an online sample has been a cause of concern, this study made a dedicated effort to minimize the limitation as data was collected throughout the year (2017). In addition to making the survey available at the Arizona State Parks and Trails website, the link was forwarded to OHV event attendees, different mailing lists, associated partners/ organizations, and OHV clubs and organized groups.

The vehicle purchase item in the spending questionnaire generated some outliers that could not be eliminated even after treating the data. Winsorized averages (obtained by trimming both ends of the data by 4.5%) are used but some outliers with low and high values could not be deleted. A future study may add a question for further clarification if a visitor reports low expenditure for the vehicle purchase item. Furthermore, it was not possible to capture the vehicle-related spending of out-of-state visitors since it was assumed that most expenditures would occur in the place of residence. It is likely that some portion of vehicle-related expenses are incurred in the visited region. A future study could include a question that could help capture that portion of vehicle-related expenditures. Because of this limitation, it is likely that the economic impact of vehicle-related spending is underestimated.

Last, this study is only able to present descriptive results from the onsite survey data. Although most of the spending figures from onsite surveys were comparable to those reported by the web-based surveys, the sample size was small and could not be included in the economic impact analysis. Still, this study has an interesting point in that it also includes retained spending of locals residing in the study area for each trail site. It is argued that spending by residents, who would have gone to a substitute OHV recreation trail outside the state if the OHV trails had not been available, is retained spending. This spending is retained by the state and its absence would have made the local resident incur spending outside the study area.

References

Arizona State Parks (2018). Off-highway vehicle recreation report. Economic impact of off-highway recreation in the state of Arizona. Retrieved from: https://d2umhuun-wbec1r.cloudfront.net/gallery/0004/0052/4CD7751111BA46518D5CBDE85C71493A/Final%20Report.pdf.

Bosworth, D. (2004). *Four threats to the national forests and grasslands. Idaho environmental forum.* Boise, ID: USDA Forest Service.

Bowker, J. M., Askew, A. E., Cordell, H. K., Betz, C. J., Zarnoch, S. J. & Seymour, L. (2012). Outdoor recreation participation in the United States – projections to 2060: A technical document supporting the Forest Service 2010 RPA assessment. Boise, ID: General Technical Report-Southern Research Station, USDA Forest Service (SRS-160)

Budruk, M., Andereck, K., Prateek, G. & Steffey, E. (2014). *2013–2014 Arizona State Parks trails study: Technical report.* Arizona State University (Report prepared for Arizona State Parks).

Bureau of Economic Analysis (2022). Outdoor recreation satellite account, U.S. and states, 2021. *U.S. Department of Commerce.* Retrieved from: https://www.bea.gov/news/2022/outdoor-recreation-satellite-account-us-and-states-2021.

Center for Business and Economic Research (2006). *The economic impact of the Hatfield-McCoy trail system in West Virginia.* Huntington, WV: Marshall University, pp. 1–65. Retrieved from: https://www.marshall.edu/cber/files/2021/04/2006_10_31_Hatfield-McCoy-Study.pdf.

Cordell, H. K., Betz, C. J., Green, G. T., & Stephens, B. (2008). *Off-highway vehicle recreation in the United States and its regions and states: An update national report from the national survey on recreation and the environment (NSRE)*. Retrieved from: http://warnell.forestry.uga.edu/nrrt/NRSE/IRJSRecIIrisRecIrpt.pdf.

Cordell, H., Betz, C., Green, G. & Owens, M. (2005). *Off-highway vehicle recreation in the United States, regions and states: A national report for the National Survey on Recreation and the Environment (NSRE)*. Retrieved from: https://www.fs.usda.gov/recreation/programs/ohv/final.pdf.

Crompton, J. (2010). *Measuring the economic impact of park and recreation services*. Virginia: National Park and Recreation Association.

Duval, D., Frisvold, G. & Bickel, A. (2020). The economic value of trails in Arizona – A travel cost method study. The University of Arizona: The Department of Agricultural and Resource Economics. Retrieved from: https://arizona-content.usedirect.com/storage/pages/20220628055837AZ%20Trails%20Economic%20Value_Full%20Report_3-30-2020_FINAL.pdf.

EPA (2019). Recreation economy for rural communities. EPA Smart Growth Program. Retrieved from: https://www.epa.gov/smartgrowth/recreation-economy-rural-communities#2019.

Greenwood, J., & Vick, C. (2008). *Economic contribution of visitors to North Carolina State Parks*. Raleigh, NC: North Carolina State University.

Gunthar, P., Parr, K., Graziano, M. & Carstensen, F. (2011). *The economic impact of state parks, forests and natural resources under the management of Department of Environmental Protection*. University of Connecticut, CT: Connecticut Center for Economic Analysis.

Holmes, T. P., & Englin, J. E. (2010). Preference heterogeneity in a count data model of demand for off-highway vehicle recreation. *Agricultural and Resource Economics Review*, 39(1), 75–88.

Hughes, C. A., & Paveglio, T. B. (2019). Managing the St. Anthony Sand Dunes: Rural resident support for off-road vehicle recreation development. *Journal of Outdoor Recreation and Tourism*, 25, 57–65.

Hughes, M. D., Beeco, J. A., Hallo, J. C. & Norman, W. C. (2014). Diversifying rural economies with natural resources: The difference between local and regional OHV trail destinations. *Journal of Rural and Community Development*, 9(2), 149–167.

IMPLAN (2014). *Principles of impact analysis and IMPLAN application*. Huntersville, NC: IMPLAN Group, LLC.

Jakus, P. M., Keith, J. E., Liu, L. & Blahna, D. (2010). The welfare effects of restricting off-highway vehicle access to public lands. *Agricultural and Resource Economics Review*, 39(1), 89–100.

Kaliszewski, N. (2011). *Jackson Hole Trails Project economic impact study*. Laramie, WY: University of Wyoming.

Kil, N., Holland, S. M. & Stein, T. V. (2012). Identifying differences between off-highway vehicle (OHV) and non-OHV user groups for recreation resource planning. *Environmental Management*, 50(3), 365–380.

Legg, M., Price, J. & Williams, P. (2006). *An economic impact analysis of the effects to Gilmer, Texas and local surrounding communities of Barnwell Mountain Off-Highway Vehicle Recreation Area, 2005–2006*. Austin, TX: Impact DataSource.

Magnini, V., Lindsey, S. & Wyatt, C. (2022). Strategically positioning Europe as a destination for off-highway vehicle recreationists: Some initial findings. *European Journal of Tourism Research*, 31, 3115–3115.

Monz, C. A., Cole, D. N., Leung, Y. F. & Marion, J. L. (2010). Sustaining visitor use in protected areas: Future opportunities in recreation ecology research based on the USA experience. *Environmental Management, 45*, 551–562.

Otto, D. (2008). *The economic impact of off-highway vehicles in Iowa.* Des Moines, IA: Strategies Economic Group.

Pardue, E., and Shand, J. (2014). *The economic impact and fiscal impact of the Hatfield-McCoy Trail System in West Virginia.* Huntington, WV: Center for Business and Economic Research.

Silberman, J., & Andereck, K. (2006). The economic value of off-highway vehicle recreation. *Journal of Leisure Research, 38*(2), 208–223.

Stynes, D. (2010). *Economic benefits to local communities from national park visitation and payroll, 2010.* East Lansing, MI: Michigan State University.

Taylor, D., Nagler, A., Bastian, C. & Foulke, T. (2013). *The economic impact of non-motorized trail usage on national forests in Wyoming.* Laramie, WY: University of Wyoming.

Thomas, F., Bastian, C. T., Taylor, D. T., Coupal, R. H & Olson, D. (2008). Off-road vehicle recreation in the west: Implications of a Wyoming analysis. In *Western Economics Forum, 7*(1837–2016–151774), 1–11.

UTV Action. (2019, 24 June). UTVs are an economic benefit. Retrieved from: https://utvactionmag.com/behind-the-wheel-utvs-are-an-economic-benefit/.

UTV Action. (2022, 6 December). UTVS/ATVS are America's #4 "conventional outdoor recreation". Retrieved from: https://utvactionmag.com/utvs-atvs-are-americas-4-conventional-outdoor-recreation/.

White, E., Bowker, J., Askew, A., Langner, L., Arnold, R & English, D. (2016). *Federal outdoor recreation trends: Effects on economic opportunity.* Boise, ID: USDA Forest Service. Retrieved from: https://www.fs.usda.gov/research/docs/outdoor-recreation/ficor_2014_rec_trends_economic_opportunities.pdf.

13 Interventions in the Tourism Market

Challenges, Opportunities and the Moral Limits of the Tourism Market

Can-Seng Ooi and Alberte Tøttenborg

Introduction

This chapter focuses on interventions in the tourism market. In the industry, there are various markets that make it possible for tourists to visit and enjoy foreign land; for businesses to serve their customers by providing goods and services; for firms to manage emotional ties with visitors and community members, such as reputation, brand and goodwill; and for workers and employers to negotiate and serve the industry. Therefore, when this chapter refers to the "tourism market", the term denotes more accurately an assemblage of markets and market mechanisms that serve different parts of the industry. This chapter will show that interventions in the market are inevitable, because market mechanisms are defined, shaped and institutionalized by humans (North, 1991; Roth, 2015).

Markets are also used to bring about balanced and sensitive tourism development. Researchers, policymakers and tourism practitioners are aware of tensions between industry and the community. For instance, communities are angered by overcrowding (Milano et al., 2019; Nepal & Nepal, 2019), inflation (Vinogradov et al., 2020), touristification of culture and heritage (Lai & Ooi, 2015), environmental impact (Budeanu et al., 2016), and exploitation of workers (Shelley et al., 2021). These challenges and their solutions inevitably involve some interventions and reshaping of the tourism market, whether the market is freewheeling or more planned and directed.

By focusing on interventions in the tourism market, this chapter reviews various approaches to balancing the contrasting needs of industry, community and the environment. As expected, different stakeholders attempt to shape and/or game the market system to further their own agendas and interests. In drawing attention to different stakeholder interests, we have broadly and heuristically framed the different market intervention approaches as: (1) market force-driven, (2) business-driven, (3) state-driven and (4) community-driven (see Freeman, 2010; Voegtlin & Pless, 2014; Zapata & Hall, 2012). These different market forms, in the context of this chapter, are meant to help manage and balance the diverse needs of business, people and the planet (Ooi, 2023).

To draw comparative lessons from the different tourism market interventions, we will provide examples from Tasmania. This southernmost island state of

DOI: 10.4324/9781003449027-16

Australia has a population of about half a million, offers many nature-based attractions and faces issues arising from strong tourism growth (Ooi & Hardy, 2020). The more elaborate examples from Tasmania form the empirical base to compare (see Table 13.1), and also to point out that the different market interventions are found in the same destination.

After presenting the four market forms with examples from Tasmania (and supported by brief examples from other places), we argue that the quest for balancing the needs of industry and the community in tourism will continue to be inadequate unless we address two fundamental issues. These two fundamental issues are the moral limits of the market (Ooi, 2022; Sandel, 2012; Skidelsky & Skidelsky, 2015; Storr & Choi, 2019). As explained later, the first moral limit of the market is that repugnant transactions are inevitably found in all the market forms, whether they are market force-driven, business-driven, state-driven or community-driven. The second moral limit is that the market generally distributes welfare and benefits to those who can afford to participate in market exchanges, rather than to those who need those benefits. Subsequently, we discuss the implications and lessons learned.

Tourism Development and Market Interventions

The tourism industry is constituted by different stakeholders, including visitors, tourism businesses, regulators, destination management organisations and local communities. All of them should have a say in how a destination is developed but their interests and agendas can be contrasting, if not contradicting (Gerke et al., 2023). Balancing the needs of stakeholders is necessary for tourism to be more sustainable. There are many models and mechanisms to bring these diverse interests together, and how each is chosen and used depends on many interrelated factors, including the nature of the tourism activities, the social-political context in the community, the size of the industry, the way tourism employees are organised, availability of alternative economic activities, or more generally the political economy of tourism in the destination (Bianchi, 2018; Denny et al., 2019; Mosedale, 2016). As mentioned, and as a heuristic to examine interventions of the tourism market, we have broadly identified four approaches: (1) market force-driven, (2) business-driven, (3) state-driven and (4) community-driven. The ways that these markets are shaped, managed and institutionalised differ, but they all are supposed to serve the needs of industry, community and the environment. Each of their successes is debatable, if not contested.

Market Force-Driven: Freewheeling Market and the Humane Society

Every stakeholder – businesses, visitors, residents, policymakers – can be expected to use the market as a mechanism to drive tourism development. A more freewheeling and market force-driven approach is considered attractive because it is supposedly self-regulating and devoid of vested interests of specific stakeholders; it is even considered the most "humane" approach for a society to develop (Röpke, 2014). As a feature in these free market forces, visitors will be attracted to well-managed destinations, and they will stop going to destinations that have been overtouristified and overcrowded; and if businesses and communities fear that they will lose customers, they will respond to the desires of the market and make changes to ensure that their attractions and services remain attractive. Yet, such a market force-driven approach is blamed for mass tourism, the pursuit of short-term interests by businesses, overcommercialisation, and it risks destroying a community before the market rights itself (Bianchi, 2018). Also, visitors may not have all the information to compare destinations and help them choose where to visit.

For example, the case of cruise ship tourism in Tasmania illustrates the impact of a largely market force-driven approach. Cruise ships are welcomed by many businesses, such as the stalls in the popular Saturday Salamanca Markets in Hobart, and tour companies transporting cruise tourists beyond the port area. However, with the increasing popularity of the island, there are concerns over community and social costs (Hardy, 2020). The market forces may not be aligned with the needs of the community. That was particularly felt when the first ship docked in Hobart since the start of the COVID-19 pandemic; it carried 3,000 passengers. Before its arrival, Tasmanians spent almost two years behind closed borders and were getting used to a tourist-light situation. After reopening the market, Tasmania received 150 cruise ships in the 2022/2023 season, with Hobart alone hosting 78 ships. On cruise ship days, Hobart's infrastructure was pushed to its limits, with traffic congestion and overcrowding. After the pandemic, the tourism market reopened. Without any market or health restrictions, and by serving pent-up visitor demands, sections of the Tasmanian community became overwhelmed and angry.

This example shows how the market force-driven approach lets demand and supply drive the industry. The outcomes did not account for or reflect the many interests of the community. It would take a long while more before angry residents would discourage people from visiting Tasmania. In the market force-driven model, businesses would first and foremost look after their profits. This may mean offering tourists services and products that may adversely affect the community. It is assumed that market mechanisms will take into account the welfare of the community. Milton Friedman's axiom that the business of business is business has come to define today's neoliberal capitalist state (Harvey, 2005). Businesses are not equipped and do not have competence in community development. Asking tourism businesses to look after community development and public services is like expecting a vacuum cleaner to function like a refrigerator. A more freewheeling market framework will not necessarily ensure that social and environmental needs of the community are factored into the industry.

Business-Driven Initiatives to "Game" the Market

Arguably, the second approach can be said to be in response to the market force-driven approach. While an unfettered market is welcomed by many tourism businesses, owners and operators are also keenly aware of their responsibilities and social license in the community. Market forces may not be responsive enough to community and environmental needs. So, besides contributing to jobs, taxes and economic activities, many take initiative and seek to demonstrate that they are contributing more directly to society and the environment via their corporate social responsibility (CSR) initiatives. Addressing social pressure, many operators lead ways to support their tourism-affected community, while they grow the industry. In the context of the market, CSR enhances the products and services by leveraging the good things that the firm does for the community and environment. Firms still use the market to sell their offerings, and often at a better price because of their elevated reputation or their enhanced social license (Voegtlin & Pless, 2014). Many companies in controversial industries (e.g., gambling, alcohol, weapons) are engaged in CSR (Jo & Na, 2012; Lee et al., 2013).

A popular way for businesses to operationalise CSR is through the triple bottom line (TBL). TBL is based on Freeman's stakeholder framework (Freeman, 2010; Tøttenborg et al., 2022a). Freeman argues that different aspects of society work together and that different stakeholders – industry, workers, residents, civil society, environmentalists – are intertwined. All their voices must be heard, and their needs and interests considered (Cole, 2014; Ooi & Strandgaard Pedersen, 2010). TBL formalises and helps measure the contribution of businesses to the community and the environment (Wise, 2016). Besides the economic and profit bottom line, companies must also account for what they have given to the people and community, and the planet and environment. It is thus not surprising to see that major tourism businesses boast of and advertise their efforts in protecting the environment and engaging with the community through selected projects like tree-planting schemes, closing the gender pay gap and sponsoring local sporting events.

Many countries, including Australia, are popular destinations for gambling. In Tasmania, Federal Group is a dominant player in the tourism market, owning a range of tourist accommodation, including some of Tasmania's most luxurious hotels. They are, however, primarily known for their casino and gaming assets, including a wide network of "pokies" (slot machines) in hospitality establishments (Markham et al., 2017). Gambling is a significant social ill and public health issue in Tasmania, and locals lose a reported AUD 1 million (USD 700, 000) each day (Bennett, 2022). In contextualising this number, the population of Tasmania is only about half a million. Federal Group has developed a CSR strategy to contribute to the community (Federal Group, 2022). They actively support different local community projects such as Variety, the Children's Charity and Dress for Success, which empowers women to achieve economic success. Their CSR approach is complemented by a strong political lobby to maintain loose gambling regulations (Boyce, 2017). This indicates a weakness of the business-led TBL balance sheet approach. CSR is sometimes abused and used to "whitewash" the reputation of companies (Pope & Wæraas, 2016).

State-Driven Interventions in the Market

The market-driven and business-driven approaches to managing tourism have their shortcomings. As a result, more intrusive state interventions in the tourism market are common. States intervene in markets for various reasons, including setting a market infrastructure for economic development, supporting costly but promising new industries, ensuring fair competition and breaking monopolies, and (re-) distributing the benefits of the market to all (Lehne, 2006). For example, the Tasmanian government has devised a sports tourism strategy by having a local Australian Football League (AFL) team. Setting up a team and building a stadium are expensive. However, AFL is a very popular sport in the country, and fans travel to support their teams. Having a team entails building a new stadium at the waterfront of the capital city, Hobart. It would cost AUD 750 million (USD 500 million). The proposed development plans have divided Tasmanians. Some believe it to be a positive development, as it will bring interstate visitors, particularly during the quiet winter months when many tourism businesses struggle to keep business going. It should also create thousands of jobs. A Tasmanian AFL team will help rally people and contribute to community building. The project will be publicly supported and also privately funded, like in a public-private partnership. In contrast, many argue that the money can be better spent, such as in solving the housing crisis, improving the dysfunctional transport infrastructure and rescuing the crumbling healthcare system (Rowbottom, 2023).

This example alludes to how states can get the most from the market by shaping the terms of what is traded, and what services can be obtained from the private sector in a public-private partnership (PPP). Governments have a responsibility to society and the environment, and they want to tap into the expertise and financial savvy of the private sector. Many infrastructures and amenities, such as roads, parks and museums, serve both residents and visitors. Developing these public goods is costly, and the civil service lacks the expertise to build, operate or maintain them (Ruhanen, 2013). Partnering with private firms to develop public goods has become popular in tourism development (Mariani & Kylänen, 2014). The private and public sectors have complementary expertise, and a PPP project focuses on common goals and objectives. Ideally, with the commercial expertise of businesses and the competencies of the public sector in serving the people, tourism development will be effective, efficient and even profitable (Zapata & Hall, 2012). Tapping into partnerships can be effective, but can also create dangers of nepotism, favoritism and perceived or real bias (Ooi, 2023).

Government-led interventions in the markets, including setting up PPP and supporting promising-but-too-risky-for-the-private-sector tourism projects, require "picking winners". The state has the role of ensuring that the diverse interests of businesses and citizens are met, and that there is economic prosperity, social stability and environmental sustainability. State intervention and support is common in tourism, for example, gazetting and running national parks and heritage sites. More directly, for instance, Bhutan charges high visitor taxes to draw in only wealthy visitors (Nyaupane & Timothy, 2010). In this instance, the tourism market is being redesigned to influence economic behavior. The state intervenes in the market to serve and achieve certain goals. The shutting down of the tourism industry around the world during the COVID-19 pandemic shows that tourism is an economic driver that can be curtailed. Many governments considered then that public health and saving lives were more important than travels. Simultaneously, many governments also provided economic support for businesses and workers in tourism and hospitality. In other words, the tourism market can be managed, regulated and even destroyed. And just as importantly, the state has the responsibility to maintain economic prosperity, ensure social stability and wellbeing, and to distribute economic benefits to the wider society.

Community-Led, or Supply-Driven

Finally, the above approaches to a balanced and sustainable tourism development strategy are challenged by many researchers and community members because they are not primarily local community-driven. It is argued that businesses, the market and the state cannot be trusted to address the serious issues of justice, fairness and equity in society (Fennell, 2018; Jamal, 2019). The influence of the tourism and hospitality industry in politics, the outsize economic dependency on the visitor economy in many places and the touristification of local life have led to demands that local residents and communities should instead drive the tourism development agenda (Higgins-Desbiolles & Bigby, 2022). Local stakeholders that have been marginalised in the big scheme of tourism planning and residents and smaller business enterprises, for instance, should play a more significant role. Many researchers and practitioners propose that tourism development should be more ground-up and more community-led (Higgins-Desbiolles et al., 2022; Muganda et al., 2013). Local groups know their place, culture and environment better than anyone else. They will know what is considered appropriate to offer to visitors, and what is not (e.g., sacred heritage sites). Community-based tourism (CBT) has thus been championed by many (e.g. Okazaki, 2008; Sin & Minca, 2014). More recently, besides CBT, regenerative tourism has been advocated. With the community embedded in the centre, regenerative tourism aims to create a resilient and resourceful ecosystem where the community and environmental and economic systems are in balance, with no groups being exploited (Gerke et al., 2023). The benefits of tourism should be distributed fairly to the community, and not skewed towards powerful business, for example.

This is easier said than done. Derived from a consultation process that lasted more than two years and listened to hundreds of Tasmanians' stories, the Tasmanian place brand maintains to provide an honest portrayal of what it means to be Tasmanian (Brand Tasmania, 2023). Even though it is for marketing the state to attract visitors, investors and skilled workers, and to export goods and services to the world, the Tasmanian brand process was predominantly community-focused; local stakeholders – businesses and residents – were asked how they would describe their island-state, and how the brand should contribute to the social, cultural and environmental wellbeing of Tasmania and Tasmanians (Tøttenborg et al., 2022b). Their stories were distilled into the brand in a process co-designed with community members. Very importantly, the brand was to be inclusive; the idea was that Tasmanians should be able to see the brand in themselves and identify with it (Tøttenborg et al., 2022a). Yet, despite the effort, many people in Tasmania are unaware or do not care about the brand; others voice criticisms. Some members maintain that the brand is a commercial enterprise, the project benefits businesses only and their community identity should not be commodified (Tøttenborg et al., 2022a).

Market Interventions and the Moral Limits of the Market

All four tourism market intervention approaches are likely to be used in combination in a single destination. Why, when, what and how each is used demands a longer discussion. For this chapter, it suffices to highlight the common and fundamental challenges facing all these four approaches. There are at least two fundamental challenges they aim to address but with varying results. The first is how various tourism activities affect the community, the environment and its cultural fabric – what is acceptable and what is not? The second fundamental challenge is how to distribute the benefits of tourism to the community in a fair and equitable way. Answers to these challenges are unclear, and we will argue next that it remains unclear because we are dealing with the moral limits of the market (Table 13.1).

Moral Limit A: Thin Line Between Acceptable and Repugnant Transactions

Through the market, money has shaped social relations and how we view value in products and services (Simmel, 1978). As a universal means of market exchange, money has become the common way of valuing things. However, some things should not be valued in monetary terms, such as love and life. This point does not discount the many advantages of allowing us to buy countless products and services. It enhances our personal liberty and freedom. Instead of bartering, monetary exchanges lower transaction costs, and are efficient and effective (Fligstein, 2002; North, 1991; Williamson, 1998). Instead of having to build trust between persons to facilitate bartering, money allows us to make instantaneous no-fuss exchanges. Tourism is only made possible because we can travel to places without knowing hosts personally. We sleep in hotels, eat in restaurants and access experiences and attractions in strange destinations.

But money should not be able to buy everything. There are products, services and experiences that are sacred, revered and priceless. If they are exchanged monetarily, they become repugnant transactions (Roth, 2015; Sandel, 2012). For example, Uluru/Ayers Rock is sacred to the indigenous Anangu people in that area in Australia. After many decades of campaigning, that attraction can now only be appreciated from afar. Tourists are no longer permitted to climb it. The national park's board of management, where Uluru is located, closed the attraction to climbers in 2019 and a breach will carry penalties under the Environmental Protection and Biodiversity Act. There are, however, still Australians who disagree with this policy. The line between acceptable and repugnant tourism transactions is unclear; there is a spectrum of transactions that some would find repugnant while others find acceptable. The first moral limit of the market points to how economic exchange transforms products, services and/or experiences in ways that may denigrate and even destroy the intrinsic values of what is being bought. Uluru, for instance, has been denigrated. Tourism market transactions are seen to have some corrupting influences on the culture, heritage and social life (Storr & Choi, 2019). Repugnant transactions may be too lucrative and thus still occur.

Moral Limit of the Market B: Unfair Distribution of Benefits

Many visitors want to support local businesses because they want the authentic experience and to "go native" (Gerke et al., 2023). There are also complaints about revenue leakage as multinational companies own hotels, restaurants and attractions, and not enough profits are given back to the local community (Lasso & Dahles, 2021). Similarly, hospitality workers do not seem to be benefitting as much as they should. For instance, even in a developed country like Australia, data from Tasmania shows that one-third of the tourism and hospitality workforce live below the official poverty standard even though the industry is a promising economic driver (Denny et al., 2019). The tourism market, like many other markets, allows for economic transactions that are very beneficial, but the benefits are not distributed in equitable and fair ways.

Essentially, accessibility to tourism goods and services is based largely on a person's ability to pay in the market, rather than based on a person's needs. For instance, landlords find it more lucrative to rent out their properties to short-term visitors, rather than to local residents. As a result, locals in many popular destinations experience housing shortage and rent increases (Wachsmuth & Weisler, 2018). The second moral limit of the market highlights the failure of the market in distributing the benefits to those who need the products and services most, and failure to distribute the benefits from the market more equitably.

Table 13.1 maps out the four market intervention approaches to the two moral limits of the market, using Tasmanian examples to provide empirical context.

Conclusions and Implications

The four interventions focus primarily on the stakeholders and their agendas and interests. As a result, not much attention is given to how the tourism market functions. This chapter suggests that it is necessary to address the two moral limits directly, so that we can determine the lines between acceptable and repugnant transactions, and how benefits from the industry should be spread to the different parties.

The moral limits of the market identified at least three issues that have to be addressed. First, it should be ensured that repugnant transactions do not occur. Second, the benefits of tourism should be more equitably distributed. And third, it should be ensured that local residents have access to resources that tourists also demand even though locals have lower spending power. How can that be done?

The market is not evil but there are moral limits. There are transactions that are repugnant, and the benefits may not be equitably distributed. The approach then is not to replace the market but to address the shortcomings in specific instances. Table 13.2 attempts to do so by framing the challenges and explaining why the different interventions into the tourism market can be problematic.

Table 13.1 A summary of the different tourism interventions in relation to the moral limits of the market

	Examples from Tasmania	Moral limit A	Moral limit B
Market force-driven	Unfettered influx of cruise passengers visiting Tasmania; as long as there is enough operational capacity at the port, cruise ships are welcomed.	Mass numbers of cruise passengers affect the character of many places, e.g., the serene atmosphere at the top of kunanyi/Mount Wellington. Without capping numbers, a freewheeling market approach will not moderate the crowds and maintain a more serene experience on the mountain. A noisy and busy kunanyi is repugnant to many residents.	Creates inconveniences for residents and most do not benefit from the increased numbers of cruise passengers. A market force-driven approach does not necessarily mitigate these inconveniences.
Business-driven	Federal Group's CSR strategy selectively supports community initiatives while creating/exploiting a vulnerable group.	Gambling harm lies behind the massive profits of Federal Group. Gambling services that nudge punters into becoming addicts are repugnant.	While some community groups benefit from the charitable work of Federal Group, many Tasmanians suffer from gambling harm. CSR does not solve the latter.
State-driven	Building the AFL stadium is part of a community-building, sport tourism and urban regeneration programme. The initiative has many public-private partnership projects.	While many Tasmanians support a local AFL team, many also find the project repugnant because it is costly and the money could be better used on healthcare, public transportation and social housing.	Developers, the AFL owners and sports fans will benefit most from the project, that is supported heavily by taxpayers.
Community-driven	"Tasmanian" the brand was created through extensive community consultation and was co-designed and led by stakeholders. The project collected stories from the ground-up and offers diverse stories.	Some Tasmanian found the still positive and heroic stories failed to tell an accurate and authentic story of the state. Honest stories should be told and not sold because the commercialisation process bastardises those stories.	The positive and heroic spin to the many brand stories have alienated many Tasmanians, including ethnic migrant members.

Table 13.2 Addressing the moral limits of the market in the context of the different market intervention models

	Stop repugnant transactions	*Distribute benefits more equitably*	*Ensure those in need get the resources*
Market force-driven	Will be hard to stop if the price is high and the transaction lucrative. *Intervention*: Regulate certain types of transactions.	There is a tendency for those who have more resources to shape the market and industry, to further their interests and benefits. *Interventions*: Regulations, taxes and policy to ensure fairer distribution of benefits from the market. That may mean that the market is less freewheeling.	
Business-driven	Firms will try to navigate and mitigate negative responses to the harm they produce through CSR. *Interventions*: More transparency and accountability of CSR activities, and directly stop or lessen harm that transactions may create.	Businesses take on CSR to ensure that they also benefit locals. Employees and community members may demand more. *Interventions*: More transparency and accountability of CSR activities, and regulation so that resources are used to support groups that have been harmed by transactions. A "right" will not necessarily cancel out a "wrong".	
State-driven	While the public and private sectors may complement each other to bring about social and environmental good, the market mechanisms may tempt parties to collude. *Interventions*: The rule of law, transparency and accountability with extensive community consultation needed.	Market mechanisms are tweaked and policies are shaped accordingly. Dangers of wealthy donors and industry lobbies shaping the policy sphere. Industry lobbies set the agenda and narrative on what is considered fair. *Interventions*: Strengthen democratic structures by enhancing the rule of law, transparency and accountability.	
Community -driven	Community-level politics will mean local elites may define what are acceptable and what are repugnant transactions. *Interventions*: The rule of law, transparency, accountability and broad community consultation necessary.	Community political elites may promote illiberal agendas and benefit themselves. The way benefits are distributed may tie to traditional feudal ways. Those who can access resources may reflect embedded social injustices and inequity in the community. *Interventions*: Strengthen democratic structures by enhancing the rule of law, transparency and accountability.	

There are at least three broader lessons. First, extensive and continuous community consultation is needed to determine what transactions are repugnant and which are not. While residents may welcome visitors at first, they may change their minds when overwhelmed by resulting changes. Second, transparency and accountability are needed to scrutinise CSR activities, state-supported and community-supported projects. What is being presented and sold to the public may not be what they actually are. Furthermore, corruption, nepotism and cronyism will not distribute the benefits of the market fairly and equitably, even if the market activities are community-led.

We show that all different market intervention approaches have their challenges, and these challenges are embedded in the market mechanisms. This also alludes to future research opportunities. Even for CBT, for instance, local elites and the powerful may still further their own personal and familial interests, instead of working for their community-at-large (Ooi, 2023). Without confronting the moral limits of the market, the challenges of different tourism interventions will continue to present themselves (Fennell, 2002; Jamal, 2019). The interventions needed for better tourism development are then targeting the specific issues at hand (Higgins-Desbiolles and Bigby, 2022). If the arbiter is to bring about a fairer tourism and also a fairer society, then it must also have a set of democratic processes, respect for the rule of law and a strong enforcement mechanism to make sure that everyone plays fairly.

References

Bennett, M. (2022). *What's the Real Cost?* Anglicare.

Bianchi, R. (2018). The political economy of tourism development: A critical review. *Annals of Tourism Research, 70,* 88–102. https://doi.org/10.1016/j.annals.2017.08.005.

Boyce, J. (2017). *Losing Streak: How Tasmania was Gamed by the Gambling Industry.* Redback.

Brand Tasmania. (2023). The Tasmanian story. https://tasmanian.com.au/be-tasmanian/the-tasmanian-story/.

Budeanu, A., Miller, G., Moscardo, G. & Ooi, C.-S. (2016). Sustainable tourism, progress, challenges and opportunities: An introduction. *Journal of Cleaner Production, 111,* 285–294. https://doi.org/10.1016/j.jclepro.2015.10.027.

Cole, S. (2014). Tourism and water: From stakeholders to rights holders, and what tourism businesses need to do. *Journal of Sustainable Tourism, 22*(1), 89–106. https://doi.org/10.1080/09669582.2013.776062.

Denny, L., Shelley, B. & Ooi, C. (2019). Education, jobs and the political economy of tourism: Expectations and realities in the case of Tasmania. *Australasian Journal of Regional Studies, 25*(2), 282–305. https://www.anzrsai.org/assets/Uploads/PublicationChapter/AJRS-25.2-pages-282-to-305.pdf.

Federal Group (2022). Federal Group: Our community. https://www.federalgroup.com.au/community/.

Fennell, D. (2002). *Eco-tourism Programme Planning.* CABI

Fennell, D. A. (2018). *Tourism Ethics* (2nd ed.). Channel View Publications.

Fligstein, N. (2002). *The Architecture of Markets: An Economic Sociology of Twenty-First-Century Capitalist Societies.* Princeton University Press.

Freeman, R. E. (2010). *Strategic Management: A Stakeholder Approach.* Cambridge University Press.

Gerke, M., Ooi, C.-S. & Dahles, H. (2023). Bourdieu on Tasmania: How theory of practice makes sense of the emergence of regenerative tourism. In E. Çakmak, R. K. Isaac & R. Butler (Eds.), *Changing Practices of Tourism Stakeholders in COVID-19 Affected Destinations* (pp. 121–141). Channel View.

Hardy, A. (2020). Cruise shipping in Tasmania. In C. Ooi (Ed.), *Tourism in Tasmania* (pp. 31–39). Forty South.

Harvey, D. (2005). *A Brief History of Neoliberalism*. Oxford University Press.

Higgins-Desbiolles, F. & Bigby, B. C. (2022). A local turn in tourism studies. *Annals of Tourism Research, 92*, 103291. https://doi.org/10.1016/j.annals.2021.103291.

Higgins-Desbiolles, F., Scheyvens, R. A. & Bhatia, B. (2022). Decolonising tourism and development: from orphanage tourism to community empowerment in Cambodia. *Journal of Sustainable Tourism*, 1–21. https://doi.org/10.1080/09669582.2022.2039678.

Jamal, T. (2019). *Justice and Ethics in Tourism*. Routledge.

Jo, H. & Na, H. (2012). Does CSR reduce firm risk? Evidence from controversial industry sectors. *Journal of Business Ethics, 110*(4), 441–456. https://doi.org/10.1007/s10551-012-1492-2.

Lai, S. & Ooi, C.-S. (2015). Branded as a World Heritage city: The politics afterwards. *Place Branding and Public Diplomacy, 11*(4), 276–292. https://doi.org/10.1057/pb.2015.12.

Lasso, A. H. & Dahles, H. (2021). A community perspective on local ecotourism development: Lessons from Komodo National Park. *Tourism Geographies*, 1–21. https://doi.org/10.1080/14616688.2021.1953123.

Lee, C.-K., Song, H.-J., Lee, H.-M., Lee, S. & Bernhard, B. J. (2013). The impact of CSR on casino employees' organizational trust, job satisfaction, and customer orientation: An empirical examination of responsible gambling strategies. *International Journal of Hospitality Management, 33*, 406–415. https://doi.org/10.1016/j.ijhm.2012.10.011.

Lehne, R. (2006). *Government and Business* (2nd ed.). CQ Press.

Mariani, M. M. & Kylänen, M. (2014). The relevance of public-private partnerships in coopetition: Empirical evidence from the tourism sector. *International Journal of Business Environment, 6*(1), 106–125.

Markham, F., Kinder, B. & Young, M. (2017, 5 April). How one family used pokies and politics to extract a fortune from Tasmanians. *The Conversation*.

Milano, C., Cheer, J. M. & Novelli, M. (Eds.) (2019). *Overtourism: Excesses, Discontents and Measures in Travel and Tourism*. CABI.

Mosedale, J. (2016). Conclusion: Tourism and neoliberalism: States, the economy and society. In J. Mosedale (Ed.), *Neoliberalism and the Political Economy of Tourism* (pp. 157–166). Routledge.

Muganda, M., Sirima, A. & Ezra, P. M. (2013). The role of local communities in tourism development: Grassroots perspectives from Tanzania. *Journal of Human Ecology, 41*(1), 53–66. https://doi.org/10.1080/09709274.2013.11906553.

Nepal, R. & Nepal, S. K. (2019). Managing overtourism through economic taxation: Policy lessons from five countries. *Tourism Geographies*, 1–22. https://doi.org/10.1080/14616688.2019.1669070.

North, D. C. (1991). Institutions. *Journal of Economic Perspectives, 5*(1), 97–112. https://doi.org/10.1257/jep.5.1.97.

Nyaupane, G. P. & Timothy, D. J. (2010). Power, regionalism and tourism policy in Bhutan. *Annals of Tourism Research, 37*(4), 969–988. http://www.sciencedirect.com/science/article/pii/S016073831000037X.

Okazaki, E. (2008). A community-based tourism model: Its conception and use. *Journal of Sustainable Tourism, 16*(5), 511–529. https://doi.org/10.1080/09669580802159594.

Ooi, C. (2022). Sustainable tourism and the moral limits of the market: Can Asia offer better alternatives? In A. S. Balasingam & Y. May (Eds.), *Asian Tourism Sustainability* (pp. 177–197). Springer Nature.

Ooi, C.-S. (2023). The local turn in tourism: Place-based realities, dangers and opportunities. In F. Higgins-Desbiolles & B. C. Bigby (Eds.), *The Local Turn in Tourism: Empowering Communities* (pp. 113–127). Channel View.

Ooi, C. & Hardy, A. (Eds.) (2020). *Tourism in Tasmania*. Forty South.

Ooi, C.-S. & Strandgaard Pedersen, J. (2010). City branding and film festivals: Re-evaluating stakeholders' relations. *Place Branding and Public Diplomacy*, 6(4), 316–332.

Pope, S. & Wæraas, A. (2016). CSR-washing is rare: A conceptual framework, literature review, and critique. *Journal of Business Ethics*, 137(1), 173–193. https://doi.org/10.1007/s10551-015-2546-z.

Röpke, W. (2014). *A Humane Economy: The Social Framework of the Free Market*. Intercollegiate Studies Institute.

Roth, A. E. (2015). *Who Gets What and Why: The Hidden World of Matchmaking and Market Design*. William Collins.

Rowbottom, C. (2023, 13 May). No stadium, no team, no deal. Will Tasmania's AFL stadium survive a minority government bombshell? *ABC News*.

Ruhanen, L. (2013). Local government: Facilitator or inhibitor of sustainable tourism development? *Journal of Sustainable Tourism*, 21(1), 80–98. https://doi.org/10.1080/09669582.2012.680463.

Sandel, M. J. (2012). *What Money Can't Buy: The Moral Limits of Markets*. Farrar, Straus and Giroux.

Shelley, B., Ooi, C.-S. & Denny, L. (2021). The dialogic negotiation of justice. *Journal of Sustainable Tourism*, 29(2–3), 488–502. https://doi.org/10.1080/09669582.2020.1727487.

Simmel, G. (1978). *The Philosophy of Money*. Routledge & Kegan Paul.

Sin, H. L. & Minca, C. (2014). Touring responsibility: The trouble with 'going local' in community-based tourism in Thailand. *Geoforum*, 51, 96–106. https://doi.org/10.1016/j.geoforum.2013.10.004.

Skidelsky, E. & Skidelsky, R. (2015). The moral limits of markets. In E. Skidelsky & R. Skidelsky (Eds.), *Are Markets Moral?* (pp. 77–102). Palgrave Macmillan.

Storr, V. H. & Choi, G. S. (2019). *Do Markets Corrupt Our Morals?* Palgrave Macmillan.

Tøttenborg, A., Ooi, C.-S. & Hardy, A. (2022a). Giving and taking ownership of a destination brand: Mechanisms of stakeholder engagement. *Journal of Place Management and Development*, 15(4), 511–532. https://doi.org/10.1108/JPMD-12-2020-0124.

Tøttenborg, A., Ooi, C.-S. & Hardy, A. (2022b). Place branding through public management lenses: Supplementing the participatory agenda. *Place Branding and Public Diplomacy*, 19, 114–127. https://doi.org/10.1057/s41254-021-00252-0.

Vinogradov, E., Leick, B. & Kivedal, B. K. (2020). An agent-based modelling approach to housing market regulations and Airbnb-induced tourism. *Tourism Management*, 77, 104004. https://doi.org/10.1016/j.tourman.2019.104004.

Voegtlin, C. & Pless, N. M. (2014). Global governance: CSR and the role of the UN Global Compact. *Journal of Business Ethics*, 122(2), 179–191. https://doi.org/10.1007/s10551-014-2214-8.

Wachsmuth, D. & Weisler, A. (2018). Airbnb and the rent gap: Gentrification through the sharing economy. *Environment and Planning A: Economy and Space*, 50(6), 1147–1170. https://doi.org/10.1177/0308518X18778038.

Williamson, O. E. (1998). Transaction cost economics: How it works; where it is headed. *De Economist*, 146, 23–58. https://doi.org/10.1023/A:1003263908567.

Wise, N. (2016). Outlining triple bottom line contexts in urban tourism regeneration. *Cities*, 53, 30–34. https://doi.org/10.1016/j.cities.2016.01.003.

Zapata, M. J. & Hall, C. M. (2012). Public–private collaboration in the tourism sector: Balancing legitimacy and effectiveness in local tourism partnerships. The Spanish case. *Journal of Policy Research in Tourism, Leisure and Events*, 4(1), 61–83. https://doi.org/10.1080/19407963.2011.634069.

14 The Challenge Contains the Solution

Designing Effective Tourism Development Interventions

Martine Bakker

Introduction

Planning interventions for tourism development projects and programs that have the goal to reduce poverty and inequality within a country or region can be especially difficult since there are many factors that need to be integrated in this process. Tourism development projects thus require a comprehensive and systematic approach as they are often dealing with multiple local, national, and international public and private sector stakeholders that have varying levels of power. These stakeholders are frequently situated in complex geo-political settings, and broader economic, social, and natural environments lead to stakeholders having different predetermined ideas and agendas about which interventions need to be implemented. If infrastructure improvements can unleash the ability of tourism to generate greater development through capacity building or improvement of the current tourism products for instance, then how is this best achieved? In most cases there are not sufficient financial and human resources to fund all the desired interventions. So, the question raised is how can the different stakeholders understand which interventions will provide 'the biggest bang for their buck'?

One of the main models that is used in development projects and programs is the *Theory of Change* (ToC) (Stein & Valters, 2012). A ToC can guide a project from identifying underlying problems to determining the desired long-term outcomes by designing appropriate interventions resulting from understanding key causal relationships. The application of a ToC in tourism development projects provides an opportunity to design interventions that are more effective and can enable more sustainable and inclusive tourism. One of the most critical steps in the design of a ToC is understanding the development challenges – what is the current reality and what are the constraints that prohibit the tourism sector from being more sustainable and inclusive? This chapter first outlines the application of the ToC approach in tourism development projects. It then presents and discusses several diagnostic tools that can assist in identifying the binding constraints that need to be addressed to realize the desired outcomes of a tourism development project.

DOI: 10.4324/9781003449027-17

Using the Theory of Change in Tourism Development Projects

One way of understanding the relationship between development challenges, interventions, and outcomes is the Theory of Change. The ToC is a process-oriented approach that is one of the most recent and frequently used concepts in development thinking (Stein & Valters, 2012). The ToC is based on the program theory and defined as 'an ongoing process of reflection to explore change and how it happens – and what that means for the part we play in a particular context, sector and/or group of people' (James, 2011, p. 27). In other words, it can explain why and how an intervention works. It is an evidence-based approach to describe 'the pathways of change that lead to the long-term goal and the connections between activities, outputs, and outcomes that occur at each step along the way' (Taplin et al., 2013, p. 2). The ToC forces its users to make knowledge gaps clear and revisit them over time (Valters, 2014) while 'giving priority to local leadership and local capacity in the search for solutions to contextually identified problems' (Booth & Unsworth, 2014, p. 3). A ToC is used before the start of a program by working backwards from the desired outcomes and, thus, minimizes the number of assumptions. The ToC is widely used by governments, companies, NGOs, and development organizations to evaluate ongoing projects or programs or evaluate them ex-post.

The *Logical Framework Approach* (LFA) and the *Results Framework* (RF), developed in the 1960s and 1970s, provided the foundation for the ToC (Prinsen & Nijhof, 2015). LFAs and RFs provide an explicit articulation (graphic display, matrix, or summary) of the different levels, or chains, of results expected from a particular intervention – project, program, or development strategy (Roberts & Khattri, 2012). The main difference between the ToC and the LFA and RF is that the latter start with the intervention and connect this to an outcome. The ToC starts with the desired outcomes and an understanding of the root problems that prevent the desired outcome. The ToC provides a much broader context and forces the user to understand the connections between current challenges, interventions, and outcomes. This prevents developing activities before fully understanding the problems.

Despite being used widely, the academic tourism literature on ToC is still very limited. Phi et al. (2018) mention that ToC has the 'potential to generate

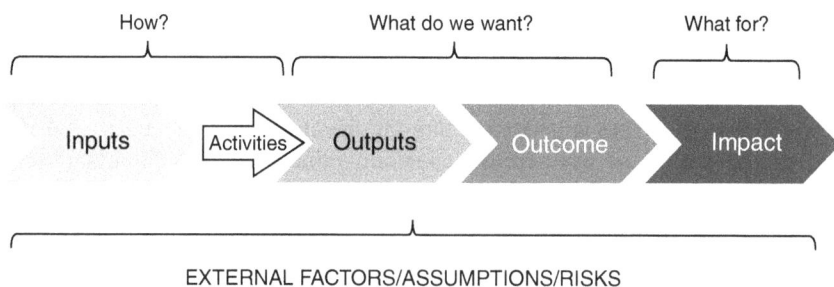

Figure 14.1 Theory of Change

Source: RVO (2018)

valuable knowledge that can lead to a well-informed debate and enhance future decision-making regarding tourism and poverty alleviation' (p. 1931). Phi et al. (2018) utilize the ToC approach in a case study on micro financing tourism development in Vietnam and argue that a ToC approach 'effectively opens up the intervention's "black box", making the assumptions and activities inside explicit' (p. 1934). Warnholtz et al. (2022) applied the ToC to evaluate the outcome of a set of interventions as part of a community-based tourism project in Mexico. They found that the design of the intervention had been too focused on the outcome without fully understanding the intervention. A study by Montano et al. (2023) applied a participatory ToC design to an EU-funded tourism project and found that the ToC is a tool that can support the evaluation of large tourism interventions in all phases of the project: planning, implementation, and closure. The stakeholders were guided through the process of developing a ToC. However, while there was some elaboration the context, the steps taken to identify the challenges were not described.

The most comprehensive progress in methodology thinking for application of the ToC in a tourism development context comes from Twining-Ward et al. (2021). They present a five-step process for developing a ToC in a tourism development project context. Their approach is based on World Bank research that outlines the use of ToC in the design and evaluation of World Bank tourism projects and programs (Twining-Ward et al., 2018). This ToC was developed to guide the design and evaluation of World Bank tourism projects and programs. It recognizes long-term and intermediate outcomes, recommends a set of tourism interventions, and identifies five categories of development challenges.

Twining-Ward et al. (2021) identify the following five steps in developing a ToC for a tourism project:

1. *Stakeholder Identification* – This process includes (i) identifying all actors, (ii) mapping their relationships, and (iii) developing measures to engage the actors during the different stages of the project.
2. *Long-term Outcomes and Assumptions* – Identifying the stakeholders' desired long-term outcomes and determining the conditions necessary for the intermediate outcomes to connect to the long-term outcomes.
3. *Challenges and Intermediate Outcomes* – Challenges are issues that need to be addressed to achieve the outcomes. The intermediate outcomes are milestones that need to be achieved to reach the long-term outcomes.
4. *Interventions* – Intervention pathways are activities that are designed to address the challenges and help achieve the intermediate and long-term outcomes.
5. *Selection of Indicators* – The selected indicators measure if the intermediate and long-term outcomes have been reached.

Under each of these five broad categories of challenges, more specific challenges are included. For example, the development challenge of 'weak private sector' can be caused by issues such as difficulty in accessing finance, lack of access to land, or poor marketing and market access (Twining-Ward et al., 2018). The World Bank

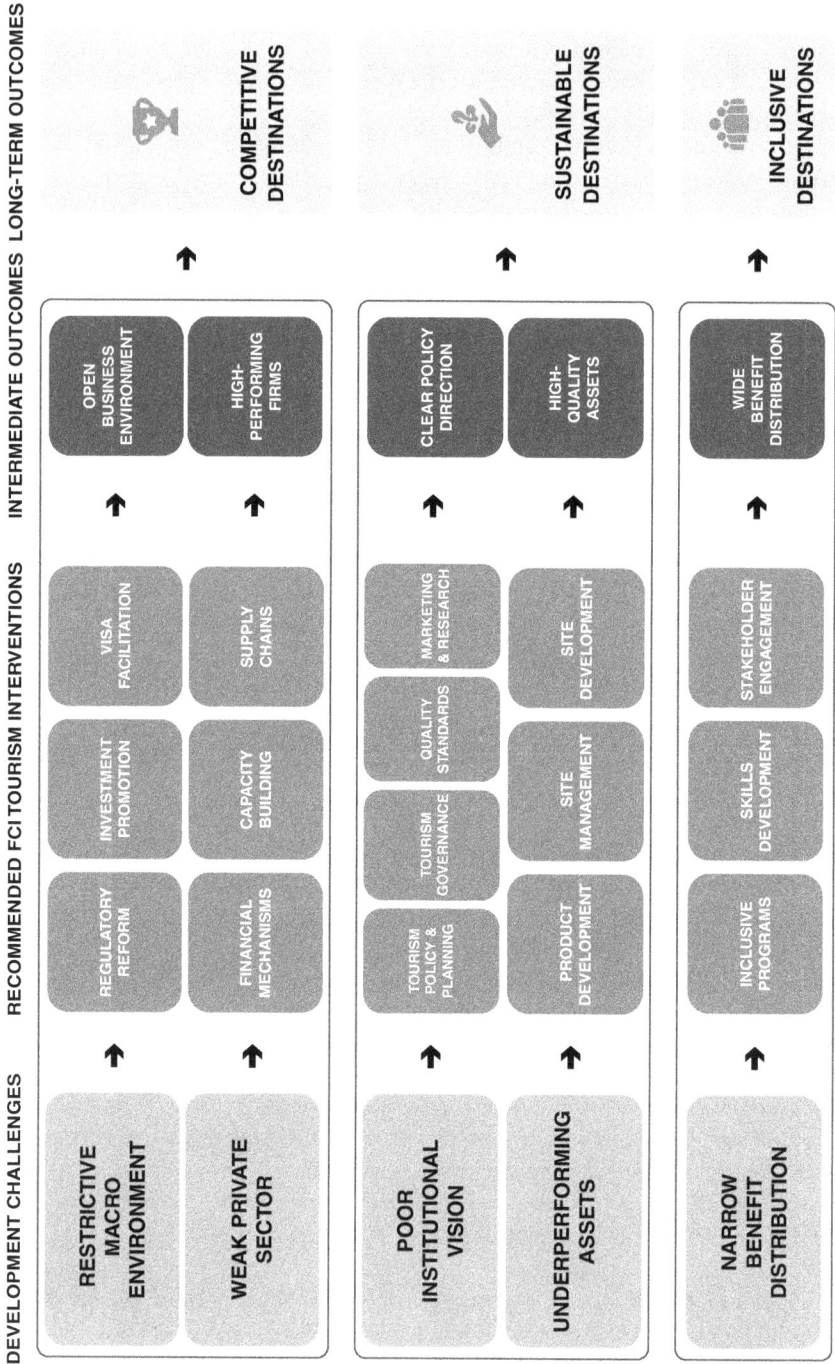

Figure 14.2 World Bank 2018 Tourism Theory of Change

Source: Twining-Ward et al. (2018)

has applied their tourism ToC to projects in Ghana, Cabo Verde, Ethiopia, and Peru among others.[1]

Tourism Diagnostic Tools

As mentioned earlier, understanding tourism challenges and constraints is one of the most critical steps in designing a tourism project. While Twining-Ward et al. (2018) provide a long list of possible development challenges, each project is unique. Consequently, a constraints analysis, in the form of a diagnostic tool, is often one of the first steps of a ToC. Warnholtz et al. (2022) argue that developing tourism as a social intervention is often happening in a context that is highly complex. As a consequence, there is a need for a methodological approach that considers 'social, economic, cultural, political and environmental interactions, influenced by interests and power struggles' (p. 2659). The authors propose the use of a methodology for designing interventions that considers the context in which the interventions take place. Twining-Ward et al. (2018) and Twining-Ward et al. (2021) also recognize this, however they do not provide a guideline or specific methodology for identifying these challenges within a specific context. There are currently several diagnostic tools specifically designed for the tourism sector that have the potential to be integrated into a ToC. Each tool has strengths and weaknesses.

World Economic Forum's Travel and Tourism Development Index

Every two years, the World Economic Forum (WEF) publishes the Travel and Tourism Development Index (TTDI). The Index was developed by the WEF in 2007 as the Travel and Tourism Competitiveness Index but transitioned in 2021 to the TTDI to include factors related to sustainability and resilience and to focus more attention on the sector's role in broader economic and social development (WEF, 2022). The 2021 Index is comprised of five subindexes, 17 pillars, 112 individual indicators, and is applied to 117 economies. The five subindexes are: (i) enabling environment, (ii) travel and tourism enabling policy and enabling conditions, (iii) infrastructure, (iv) travel and tourism demand drivers, and (v) Travel and tourism sustainability. Most of the dataset for the TTDI is comprised of statistical data from international organizations such as UNWTO, World Bank, WTTC, IATA, and others. The Index relies on survey data from the World Economic Forum's annual Executive Opinion Survey, to measure concepts that are qualitative in nature or for which internationally comparable statistics are not available for enough countries (WEF, 2022).

The Index serves as a benchmarking tool to assist countries in understanding the strengths and weaknesses of their tourism sector compared to other countries. However, while the Index highlights possible constraints, it provides limited understanding of local circumstances that guide the design of specific interventions. It is therefore a good starting point but would need to be supplemented with additional diagnostic studies when designing a complex tourism project.

Destination International's DNEXT Diagnostic Tool

The DestinationNEXT Scenario Model and Assessment Tool, developed by the consultancy firm Destination International, is designed to help tourism sector stakeholders evaluate their destinations and detect shortcomings (Destinations International, 2023). The self-assessment is based on 24 variables related to destination strength and community support and engagement. Within each variable, a series of metrics is used. Variables include metrics on destination performance, brand identity, tourism assets, access and mobility, communication and Internet infrastructure, community engagement, and more. The data is collected through a stakeholder survey which includes two parts: (i) an assessment of how important each variable is to the destination, and (ii) an assessment of the perceived performance across each of the variables. The diagnostic plots the survey results into a scenario model, which shows specific opportunities for the destination. The diagnostic report card outlines the results for each variable, and how the destination compares to the industry average. In cases where multiple stakeholders complete the assessment, a standard deviation measurement indicates how much consensus there is across the participants (Destinations International, 2023). The diagnostic has been used by smaller places such as the Rogers-Lowell area in Arkansas but also large destinations including Orlando and Hawaii.

The main strength of this approach is that the process to undertake this diagnostic can be relatively short. A weakness of this diagnostic approach is that it does not evaluate the entire landscape of elements that play a role in tourism development. The survey tool is also limited in the sense that it only measures the presence of an element and not the quality (Seakhoa-King et al., 2020). The tool can be useful as part of a larger and more comprehensive diagnostic process.

The United Nations Development Programme's Tourism Diagnostics

In 2021 and 2022, the United Nations Development Programme (UNDP) undertook a series of tourism diagnostics as part of the 'FUT-Tourism: Rethinking Tourism and MSMEs in times of COVID-19' project in five Eastern Caribbean countries. The project's goal was to provide suggestions on how to revamp the tourism sector of each country through regional dialogues as well as technical and financial assistance for micro, small, and medium enterprises (MSMEs) (UNDP, 2022). The areas of focus, as determined before the diagnostic process, were: (i) digital transformation of tourism MSMEs, (2) product diversification, and (3) stakeholder engagement and regional cooperation. The main objectives of the diagnostics were:

- To assess digital infrastructure at country level;
- To examine opportunities for digital entrepreneurship for MSMEs in the tourism sector;
- To identify market trends and opportunities for MSMEs in the tourism sector;
- To identify catalytic interventions which are gender-responsive and reflect on specific women's needs to access training, financing, or technical support; and

- To identify initiatives that will accelerate the transition towards a more inclusive and sustainable tourism industry and progress towards the Sustainable Development Goals (SDGs).

The methodology of the diagnostic studies comprised secondary and primary data collection. Secondary research included a literature review and analysis of tourism and related sector strategies and action plans, tourism demand, market assessment and performance reviews, statistical analysis, and competitive analysis. The primary data collection and consultative process included a questionnaire, focus groups with tourism MSMEs, and interviews with public and private sector tourism stakeholders. The questionnaire was designed for the Eastern Caribbean Central Bank (ECCB) to collect information about the implementation of a new regional digital currency. The MSME focus groups, and the public and private sector interviews focused only on the three specific areas that had emerged through pre-consultation with stakeholders. The studies did not investigate possible constraints concerning institutional arrangements, air transport and maritime sectors, funding mechanisms, or tourism legislation.

Data collection took place during the COVID-19 pandemic and had time restrictions. All focus groups and interviews took place virtually and did not include on-site observations. This limited the analysis as it resulted in lower-than-expected participation of MSMEs in the focus groups and limited the responsiveness of key stakeholders targeted for interviews (UNDP, 2022). The final diagnostic reports included a SWOT analysis and suggested policy interventions for each of the three focus areas for the five countries.

The main strength of this tourism sector diagnostic methodology is that it identified constraints by specific focus areas. This is also its weakness as it does not diagnose the entire spectrum of possible constraints and, therefore, cannot lead to an identification of all the interventions required. The main constraints might be found in other areas. The diagnostic also did not use a systematic approach to analyze the collected data and identify prioritized weaknesses. The diagnostic methodology as described above is a common approach for many tourism development projects.

The World Bank's Tourism Diagnostic Toolkit

In 2019, the World Bank published the *Tourism Diagnostic Toolkit* (World Bank, 2019). The toolkit provides a systematic approach for identifying and assessing opportunities and constraints in a tourism ecosystem, as well as identifying potential points of entry for interventions.

The diagnostic process includes the following steps:

1. *Project planning*
   a. Establish a team
   b. Develop a scope of work
   c. Design an implementation plan

2. *Desk research*
   a. Conduct a literature review
   b. Identify benchmarks
   c. Produce destination and stakeholder maps
3. *In-country assessment*
   a. Execute stakeholder consultation (interviews, focus groups, workshops)
   b. Site visits to assess product quality
   c. Carry out specific analysis required based on project
4. *Analysis and reporting*
   a. Analyze information
   b. Draft report and validate
   c. Finalize report

During steps two and three, the focus is on data collection reflecting: (i) context – the size of the sector and macro-setting; supply, demand, and policy and institutional context; (ii) competitive position – ranking position on WEF Index, World Bank Doing Business Index, and other relevant rankings; and benchmarking position on factors including tourism products, price competitiveness, safety and security, policy rules and regulations, human resources and tourism infrastructure; and (iii) opportunities assessment – overall country opportunity assessment, and potential policy and institutional constraints.

The diagnostic toolkit also includes an overview of common tourism constraints as observed during World Bank tourism projects and suggests approaches for addressing constraints (see Figure 14.3 for more details).

The main strength of the Tourism Diagnostic Toolkit is that it is very comprehensive and provides a clear process. However, the crucial step in this process of moving from assessment to analysis provides limited guidance on how to systematically analyze and combine the collected information to derive an understanding of the main constraints.

Tourism-Driven Inclusive Growth Diagnostic (T-DIGD)

Bakker (2019) developed a diagnostic tool, called the Tourism-Driven Inclusive Growth Diagnostic (T-DIGD) which is designed to identify the binding constraints limiting tourism-driven inclusive growth. The tool is based on the principles of the Hausmann, Rodrik, and Velasco (HRV) Growth Diagnostic, designed to investigate which factors have the most distortionary effects on a country to achieve sustainable economic growth and to determine a rational prioritization of interventions (Hausmann et al., 2008). The HRV model draws from the Theory of Constraints developed by Goldratt (1990). This theory states that an organization is a system of connected processes, rather than a collection of independent processes and there is only one 'constraint' which hinders an organization. This 'constraint' is the 'weakest link' in the system (Goldratt, 1990). Within the HRV Growth Diagnostic, these

PROBLEMATIC BUSINESS ENVIRONMENT	SUGGESTED APPROACHES
• Burdensome licensing, regulatory, and legal frameworks • Outdated and contradictory laws • Poor tourism investment environment • Ineffective incentives for tourism • Visa and access cost and burden • Government and SOEs crowding out private sector • Poor public-private dialogue (PPD) • Low security and visitor safety • Infrastructure constraints	• Regulatory reform • Investment promotion • Visa facilitation • Development of concession frameworks and other PPP mechanisms • Tendering provision of services to the private sector • Improved Infrastructure
LOW PERFORMING FIRMS	**SUGGESTED APPROACHES**
• Low product quality • Difficulty accessing finance • Closed markets for private sector • Lack of access to land • Lack of legal know-how • Poor marketing and market access • Poor working conditions • Weak SME support and incentives • Low management and business development capacity	• Quality improvement programs • Improved access to finance • Reform 'reserved' list to open markets • Opening access to public land • Land title guarantees • Capacity building • Supply chain development or strengthening • Utilization of new technology
LACK OF CLEAR POLICY DIRECTION	**SUGGESTED APPROACHES**
• Poor prioritization and vision • Lack of integrated destination planning • Lack of data on demand and supply • Limited public-sector capacity • Poor service quality • Lack of public sector investment • Poor marketing and promotion • Lack of clear direction on visa policy	• Tourism planning • Tourism policies • Tourism marketing and research • Improved standards and capacity • Licensing and regulation simplification
LOW PERFORMING ASSETS	**SUGGESTED APPROACHES**
• Poor inter-agency coordination • Poor visitor and site management • Limited income from tourism • Lack of environmental and heritage regulations • Poor heritage and culture management • Poor natural assets management • Outdated and undifferentiated product • Infrastructure constraints	• Governance and coordination • Product development • Site management and development
LOW STAKEHOLDER ENGAGEMENT	**SUGGESTED APPROACHES**
• Low female inclusion • Lack of community involvement and consultation mechanism • Weak supply chain • Low youth involvement • High geographic concentration • Low tourism awareness • High degree of informality	• Business support • Online engagement • Skills development • Stakeholder engagement • SME strengthening

Row labels (left column): COMPETITIVE CONSTRAINTS, SUSTAINABILITY CONSTRAINTS, INCLUSIVENESS CONSTRAINTS

Figure 14.3 Common tourism constraints and suggested approaches

Source: World Bank (2019)

overpowering constraints should be targeted first and are referred to as binding constraints.

> The trick is to find those areas where reform will yield the greatest return. Otherwise, policymakers are condemned to a spray-gun approach: they shoot their reform gun on as many potential targets as possible, hoping that some will turn out to be the ones they are really after (Rodrik, 2006, p. 982).

The HRV Growth Diagnostic also draws from the Theory of the Second Best as developed by Lipsey and Lancaster (1956). This theory states there is an interaction between any specific distortion and all other distortions. This means that not only the direct impact of the targeted distortion should be studied, but also the changing interrelationships between all other distortions. Consequently, the HRV Growth Diagnostic calls for a series of econometric and benchmarking tests that indicate whether a particular factor constrains growth. The diagnostic addresses binding constraints to growth but does not address inclusive growth. However, the World Bank, the Asian Development Bank (ADB), the African Development Bank (AfDB), and other institutions have started to apply inclusive growth strategies and adapted the HRV model to identify the constraints to inclusiveness of growth (ADB, 2010; USAID, 2014; World Bank, 2011).

The ability of a tourism sector to be more inclusive has become an important focal point for policymakers and scholars alike (Bakker & Messerli, 2017). The T-DIGD is based on the HRV Diagnostic but adapted to diagnose tourism sector-specific growth constraints while also diagnosing the constraints to the inclusiveness of the sector.

Figure 14.4 shows that the diagnostic categorizes the main constraints to tourism-driven inclusive growth under three pillars. Pillar I, 'Growth of tourism opportunities', predominantly takes the perspective of firms and it is strongly

Inclusive growth

Inclusive and Productive Tourism Employment Opportunities

Pillar I: Growth of tourism opportunities	Pillar II: Equal access to tourism opportunities	Pillar III: Equal outcome of tourism opportunies
• Human resource capacity • Infrastructure • Safety, political stability, security and health • Accessibility • Business policy environment • Land and property rights • Government prioritization • Environmental quality • Market coordination and responsiveness • Access to finance	• Access to education • Access to infrastructure • Access to finance • Access to land • Access to information and knowledge • Access to regulatory and political system	• Monetary • Non-monetary

Figure 14.4 Tourism-driven inclusive growth diagnostic (T-DIGD)

Source: Bakker (2019)

linked to the competitiveness of a destination. Critical success factors in determining competitiveness entail improving the business enabling environment and strengthening of the private sector. Pillar II, 'Equal access to tourism opportunities', addresses employability or access to entrepreneurial opportunities for individuals or specific groups of individuals (e.g., women, ethnic minorities, or rural communities). This second pillar examines what prohibits these groups from gaining access to productive opportunities generated by the tourism sector. Pillar III, 'Equal outcome of tourism opportunities', examines if the outcome of participating in productive employment is equal among individuals or groups of individuals and examines monetary (wages) and non-monetary outcomes for different groups.

The three pillars are closely linked to each other. Both Pillars I and II analyze factors concerning infrastructure, education, access to land and finance. This means that, for example, under Pillar I the overall tourism-related education in the country is analyzed, while under Pillar II the access by specific groups within society is analyzed. This denotes that the effects of these constraints to growth of tourism opportunities, and access to these opportunities, need to be analyzed jointly as, for example, increased access to education can create continuous circles of increased opportunity.

The methodology of applying the T-DIGD requires the following steps:

1. Document analysis, including national and regional tourism plans as well as other relevant policy documents and diagnostics, to identify an initial list of constraints to the tourism sector's growth and inclusiveness in the country under study.
2. Benchmarking the country against a competitive set using 96 indicators that measure each of the 18 factors across the three pillars.
3. Interviews with public and private sector stakeholders.
4. Site visits to observe the quality of the tourism product and infrastructure.
5. Analysis of the different datasets under each of the 18 factors, to identify the binding constraints. This includes:

   a. Examining the country compared to the benchmark countries by measuring the distance to the mean score of each indicator. An indicator score within a 90th percentile score of the mean for the benchmark countries would qualify a constraint as non-binding; between 90th and 80th percentile as moderately binding; and any score below the 80th percentile from the mean qualifies a constraint as possibly binding.
   b. The qualitative data collected through stakeholder interviews, document analysis, and site visits are used to validate the scoring.
   c. When there is no quantitative data available, the analysis is fully dependent on the available qualitative data. In cases where the qualitative data does not support the quantitative data, the information obtained through the qualitative data is weighted over that derived from the quantitative indicators.

6. The results of the three-step methodology are presented in a heat map that indicates for each of the 18 factors any challenges that can be considered either

a binding constraint, a moderately binding constraint, or that no other binding constraints are identified at that moment. The factors under Pillars II and III are disaggregated (gender, location, age, ethnicity, and socio-economic status) in order to diagnose constraints that are affecting some groups of people more than others.

The T-DIGD has been applied to the tourism sector of North Macedonia (see Box 14.1). Subsequently, Cocker (2023) applied the T-DIGD to the budget travel industry in Sri Lanka (see Box 14.2) and made an amendment to the methodology used in step 5. He added normalization of all benchmark indicator data prior to comparison via a min-max formula. This allows country comparisons to be better contextualized amongst the entire range of values in any given dataset.

Box 14.1   Application of the TDIGD to the Tourism Sector of North Macedonia

Applying the T-DIGD in North Macedonia identified several constraints deriving from the country's geographic, political, and socio-economic realities that hinder the sector's capacity to create inclusive employment opportunities. The results suggest that the main barriers for the tourism sector's ability to contribute to inclusive growth in in the country are related to the quality and access to education; the quality of the road infrastructure, especially in the rural areas; issues around political instability and safety; tourism prioritization; and finally, a lack of entrepreneurship and market responsiveness. Analysis of the constraints provided insights that could lead to the design of specific interventions. For example, the general education indicators and the presence of vocational schools and a university that offer tourism programs did not indicate that human resources could be a constraint. However, interviews with tour operators indicated difficulties in hiring qualified and skilled people. They specified that this is most visible in guiding where there is currently a lack of properly trained and skilled guides. Tour operators indicated a high demand for guides of international standard who possess the required technical, social, and language skills, and that current guide training courses do not adequately address the demand of the market.

Source: Bakker et al. (2023)

The main strength of the T-DIGD is that it provides a systematic approach through its framework that enforces a step-by-step process of considering all factors that can constrain the tourism sector from contributing to inclusive growth. It forces users to keep an open mind instead of following pre-determined opinions. The main weakness is the reliance on the availability of indicator data since limited or insufficient data hampers the process of prioritizing the constraints. While not originally designed to do so, the T-DIGD could be integrated into a ToC and used as a tool to diagnose the most critical development challenges of a tourism sector. The T-DIGD is designed to identify the most binding constraints. If these constraints are addressed through appropriate interventions, they should then also result in the most optimized use of resources.

Box 14.2   Application of the T-DIGD to the Budget Travel Industry of Sri-Lanka

Application of the T-DIGD to the budget travel industry of Sri Lanka showed that it is constrained from the production of inclusive growth through an array of factors that are specific to the needs of the industry, specific to the needs of social groupings, and common across Sri Lankan society. The diagnostic identified widespread issues with access to finance as a binding constraint, alongside restraints concerning access to land, and root cause discriminations related to distrust and specific to marginalized groups. The business policy environment for budget travel entrepreneurs was also diagnosed as a binding constraint. Foreign ownership patterns are also considered a constraint on inclusive growth, though benefits from foreign involvement in the industry are also noted. The diagnostic was able to identify specific challenges that could lead to the design of targeted interventions. For example, for constraints regarding accessibility the following was noted. The air connectivity of Sri Lanka is limited compared to other comparable destinations in the region. There are just a few small budget airlines connecting Sri Lanka with major hubs, and airport taxes and fees are also significantly higher than benchmark countries. Sri Lanka also has no visa-free travel arrangements with any other country. As such, with the exception of a few Asian countries whose residents are able to pay a $20 visa fee, visitors are paying $110 in government fees just to visit Sri Lanka, which is a disincentive for budget travelers.

The diagnostic process was hindered by a considerable lack of appropriate quantitative data to triangulate the qualitative findings. Another issue was the selection of benchmark countries for the quantitative analysis. Sri Lanka's size, geography, maturity of the tourism sector, the state of the overall economy, and its diverse culture made it challenging to select appropriate benchmark countries in the region.

Source: Cocker (2023)

Conclusion

Tourism development projects are often complex and there are many different internal and external factors at play that affect the outcome and impact of a project. The ToC approach was developed to provide a system to work backwards from desired outcomes. It can help with understanding the development challenges and aid in designing interventions that lead to the desired impact. In a tourism project, stakeholders grapple with complexity, consider the multiple issues related to tourism development, and identify interventions that can lead to 'solutions'. There are currently several diagnostic tools that can be used with the ToC to identify the main challenges that need to be addressed in determining how tourism can contribute to the development of a product, place, or country. Each diagnostic has strengths and weaknesses, and the choice of which to use is based on the available resources and the size and complexity of the project. Thus, instead of pulling interventions out of a 'black box' or using a 'spray-gun' approach, as some scholars refer to it, meaningful interventions for tourism development analysis and projects require a clear understanding of why there is need for desired change and to what it should lead. This will allow the selection of those interventions, within the boundaries of the available resources, that can lead to real change and impact.

Note

1 For most of the projects, detailed information is available at: https://projects.worldbank.org/en/projects-operations/projects-home.

References

ADB (2010). *Indonesia critical development constraints*. Asian Development Bank.

Bakker, M. (2019). A conceptual framework for identifying the binding constraints to tourism-driven inclusive growth. *Tourism Planning & Development, 16*(5), 575–590.

Bakker, M., & Messerli, H. R. (2017). Inclusive growth versus pro-poor growth: Implications for tourism development. *Tourism and Hospitality Research, 17*(4), 384–391.

Bakker, M., van der Duim, R., Peters, K., & Klomp, J. (2023). Tourism and inclusive growth: Evaluating a diagnostic framework. *Tourism Planning & Development, 20*(3), 416–439.

Booth, D., & Unsworth, S. (2014). *Politically smart, locally led development*. Overseas Development Institute.

Cocker, T. (2023). The barriers to inclusive growth in Sri Lanka's budget travel industry unlocking local, latent development potential. Master's thesis. University of Amsterdam.

Destinations International (2023). *DNEXT Diagnostic Tool*. https://destinationsinternational.org/dnext-diagnostic-tool-old.

Goldratt, E. M. (1990). *Theory of constraints*. North River.

Hausmann, R., Rodrik, D., & Velasco, A. (2008). *Growth diagnostics*. In Serra, N., & Stiglitz, J. E. (Eds.), *The Washington consensus reconsidered: Towards a new global governance* (pp. 324–355). Oxford University Press.

James, C. (2011, 11 September). *Theory of change review*. Comic Relief.

Lipsey, R. G., & Lancaster, K. (1956). The general theory of second best. *The Review of Economic Studies, 24*(1), 11–32.

Montano, L. J., Font, X., Elsenbroich, C., & Ribeiro, M. A. (2023). Co-learning through participatory evaluation: An example using Theory of Change in a large-scale EU-funded tourism intervention. *Journal of Sustainable Tourism*, 1–20.

Phi, G. T., Whitford, M., & Reid, S. (2018). What's in the black box? Evaluating anti-poverty tourism interventions utilizing theory of change. *Current Issues in Tourism, 21*(17), 1930–1945.

Prinsen, G., & Nijhof, S. (2015). Between logframes and theory of change: Reviewing debates and a practical experience. *Development in Practice, 25*(2), 234–246.

Roberts, D., & Khattri, N. (2012). *Designing a results framework for achieving results.* World Bank.

Rodrik, D. (2006). Goodbye Washington consensus, hello Washington confusion? A review of the World Bank's economic growth in the 1990s: Learning from a decade of reform. *Journal of Economic Literature, 44*(4), 973–987.

RVO (2018). *Fund Against Child Labour (FBK) – Developing your Theory of Change.* Netherlands Enterprise Agency. https://english.rvo.nl/sites/default/files/2018/11/FBK_ theory_of_change_guidelines_0.pdf.

Seakhoa-King, A., Augustyn, M. M., & Mason, P. (2020). *Tourism destination quality: Attributes and dimensions.* Emerald Publishing Limited.

Stein, D., & Valters, C. (2012). *Understanding theory of change in international development.* JSRP and TAF collaborative project. JSRP Paper 1. Justice and Security Research Program, International Development Department, London School of Economics and Political Science, London, UK.

Taplin, D. H., Clark, H., Collins, E., & Colby, D. C. (2013). Theory of change. In *Technical papers: A series of papers to support development of theories of change based on practice in the field.* ActKnowledge.

Twining-Ward, L., Messerli, H., Sharma, A., & Villascusa, J. M. (2018). *Tourism Theory of Change.* Tourism for Development Knowledge Series. World Bank.

Twining-Ward, L., Messerli, H., Villascusa, J. M., & Sharma, A. (2021). Tourism Theory of Change: A tool for planners and developers. In Spenceley, A. (Ed.), *Handbook for sustainable tourism practitioners: The essential toolbox* (pp. 13–31). Edward Elgar.

UNDP (2022). *Future tourism: Rethinking tourism and MSMES in times of COVID-19: Tourism Diagnostic Report – Eastern Caribbean.* United Nations Development Programme. https://www.undp.org/sites/g/files/zskgke326/files/2022-09/undp_diagnostic_report_ 2022_-_subregional-final.pdf.

USAID (2014). *Bangladesh Inclusive Growth Diagnostic.* United States Agency for International Development.

Valters, C. (2014). Theories of change in international development: Communication, learning, or accountability. The Justice and Security Research Program. Paper 17. 1–29.

Warnholtz, G., Ormerod, N., & Cooper, C. (2022). The use of tourism as a social intervention in indigenous communities to support the conservation of natural protected areas in Mexico. *Journal of Sustainable Tourism, 30*(11), 2649–2664.

WEF (2022). *Travel & Tourism Development Index 2021: Rebuilding for a sustainable and resilient future.* World Economic Forum. https://www.weforum.org/reports/travel-and-tourism-development-index-2021/in-full/about-the-travel-tourism-development-index#1-2-data-and-methodology.

World Bank (2011). *East Java Growth Diagnostic: Identifying the constraints to inclusive growth in Indonesia's second-largest province.* World Bank.

World Bank (2019). *Tourism for development: Tourism Diagnostic Toolkit.* World Bank. https://documents1.worldbank.org/curated/en/240451562621614728/pdf/Tourism-Diagnostic-Toolkit.pdf.

15 Policy Failures, Action and Implementation Gaps, and Non-Policy in Tourism

A Critical Appraisal

Alberto Amore and C. Michael Hall

Introduction

Recent environmental, political, and economic crises have shed further light on the vulnerability of tourism across different regions and environments around the globe. From a public policy perspective, legislation, regulations, and direct interventions of national, regional, and local governments in tourism have long been used to address various social, economic, environmental, and technological issues impacting businesses and individuals (Amore, forthcoming; Hall & Jenkins, 1995). However, evidence shows that many of these interventions in both developed and developing countries reflect the proliferation of neoliberal and hyperneoliberal ideologies emphasizing economic growth and *laissez faire* policies that reinforce the role of "the market" as a governance mechanism (Duffy, 2015; Torkington et al., 2020).

Public policy in tourism predominantly consists of reactive, *ex-post* policy actions and deliberations that are, in part, a response to the inactions of governments in the first place (Dodds & Butler, 2008). The conjunctural nature of tourism policies can be particularly observed in the extemporaneous and short-term policy responses in the aftermath of natural hazards and crises, with redevelopment almost exclusively focusing on fast-track recovery of destinations (Hall et al., 2016). Even in contexts where tourism is a pivotal contributor to economic development, the fragmented nature of tourism governance steering modes often undermines the capacity of genuinely broad and inclusive collaborative policy actions and stakeholder decision-making (Hall, 1999; Farmaki et al., 2015). This can be observed on key policy aspects like environmental protection (Aall et al., 2015), heritage governance (Adie, 2019), migration (Amore, 2023), local destination governance (Zahra, 2011), and funding (Bohlin et al., 2016). For example, the recent concept note by the UNWTO (2023) put the accent on the opportunity to invest in green infrastructure in the tourism and hospitality sector. However, it remains unclear as to how infrastructural improvements of this kind and the related rebound effects effectively contribute to what already is an energy-intensive sector (Hall & Amore, 2016).

It has been widely acknowledged that legislation and regulations for tourism, at all scales, are "rarely exclusively devoted to tourism per se" (Hall, 2008, p. 14). This may be because there is no clear government authority for tourism able to

DOI: 10.4324/9781003449027-18

design and implement a cohesive policy agenda like other quintessential branches of government (Hall, 2008). Additionally, there is no one-size-fits-all government structure for national, regional, and local tourism policy and planning given that tourism-relevant policies operate across different government agencies and ministries, with most tourism-specific government bodies tending to have a narrow promotional, research, and development focus. This, in turn, can decrease the effectiveness of tourism policies and increase the chances of inter-agency policy differences over policy settings and blockages in the policy process (Hall & Jenkins, 1995; Simmons et al., 1974). Despite its economic significance, the political environment in which tourism policymaking takes place is also likely to not regard tourism as a key socio-economic policy area for national and regional development, unless there is a high degree of economic dependency on tourism, as in some small island developing states. A review of national and regional governance frameworks for tourism in OECD countries, shows how the tourism policy arena is often subsidiary to other main government portfolios or relegated to second-tier directorates and government branches with mainly marketing and policy advisory roles (OECD, 2022).

The COVID-19 global pandemic and the impacts on tourism following the lockdowns represented a potential turning point in the study of tourism policy and planning. As Higgins-Desbiolles (2021) observes, academic debate in the field saw a polarization between those advocating for a much-needed reform towards more responsible and sustainable trajectories in tourism (e.g., Brouder, 2020) and more precautionary approaches (Hall et al., 2020). Hall and Seyfi (2021) discussed the impact of COVID-19 on tourism and the potentialities of degrowth and slowing down practices in acknowledgement of the environmental and social pressures destinations have been coping with prior to the pandemic. However, the authors noted that "many of the government responses to COVID-19 are not geared towards green responses but are instead meant to reinforce business as usual or worse" (Hall & Seyfi, 2021, p. 225). The latter echoes Amore and Adie's (2021) findings across urban destinations and the prevailing rhetoric of swift recovery following the pandemic.

The following chapter provides a critical appraisal of policy issues and policy interventions with direct impacts in tourism as well as policy problems caused by unregulated and market-driven tourism policies. It does so by presenting a framework illustrating policy failures, and policy-action relationships/policy implementation gaps and non-policies building from evidence collected from destinations and contexts in developed and developing countries. The chapter then provides an overview and a rethinking of policy action in tourism in opposition to the prevailing mode of governance and public policy making.

Theory

Research focusing on tourism and public policy builds on a well-established body of studies rooted in political theory, policy analysis, political geography, and development studies. Geographical reach and scale of analysis is distributed across different types of tourist destinations (for a recent review, see Amore, forthcoming). Overall, policy studies shed light on the reach, efficacy, and fallacies of policy actions and inactions when it comes to the role of governments in the governance of tourism. At the national level, lobbying from tourism industry stakeholders has often defined the action of governments over the last decades. This was the case, for example, of Australia, with the federal government establishing a steering committee aimed at bolstering the national tourism economy so long advocated by business stakeholders (Dredge & Jenkins, 2012). In many countries, and especially those within the Westminster tradition, such as Australia, Canada, and New Zealand, national tourism bodies are typically overseen by boards that are filled with industry members and there are particularly close policy development relationships between agencies and tourism industry associations.

Studies conducted at the regional scale suggest that economic development, diversification, and competitiveness are main reasons for policy action and implementation of regional tourism strategies (Hall, 2008; Malecki, 2004). In South Africa, for instance, the mandate by the national government to regional authorities was to ensure "the greater 'geographic spread' of tourism with emphasis given to supporting tourism growth in rural areas" (Rogerson, 2015, p. 283). Such measures are reinforced by the promotional activities of regional tourism organizations as well as competition to host events and attract infrastructure developments. Interestingly, the decision to encourage regional tourism development is often taken in response to economic restructuring, arising from technological and economic change, and perceived crises. More rarely are regional tourism policy initiatives "a proactive decision as the result of an awareness of changes in the tourism business environment" (Hall, 2008, p. 104).

Government and governance for tourism at the local level tends to be characterized by forms of coalescence between local authorities and private stakeholders on destination strategy and infrastructural development for tourism (e.g., Bramwell & Meyer, 2007). The action of policymakers is often coupled with mechanisms and forms of inaction and non-decision making to keep opposing stakeholders at arm's length or to ensure that the "rules of the game" favour particular policy interpretations (Hall, 2010a). This occurs, for instance, in relation to tourism and hospitality infrastructure development in the aftermath of natural hazards (Amore et al., 2017) or in cases where those in favour of the developments reach an agreement behind closed doors and exclude other actors from the policy process or limit their capacity to influence policy settings. The latter scenario occurs on tourism policy issues that put revenue and economic return ahead of natural and environmental protection (Hall & Amore, 2020; Higgins-Desbiolles, 2011).

It is widely acknowledged that the dogmas of neoliberalism influence the actions and decisions of governments in tourism. Overt and covert forms of power further define policies that legitimize pro-growth economic agendas (Bramwell, 2011). Duffy (2015) shows how nature-based tourism policies in Thailand and Madagascar promote a neoliberal-driven rhetoric at the expenses of wildlife protection and environmental conservation. Neoliberalism also permeates national tourism policy documents and strategies in Europe. A recent review by Torkington et al. (2020)

shows how sustainability discourses in tourism are framed under neoliberalism and growth dogmas, while Hall (2019) suggests that the notion of sustainable development, as reported in the Brundtland Report, has been used by policymakers to promote and support green growth discourses in tourism that focus on efficiency rather than the absolute impacts of tourism. According to Amore and Hall (2017), crises and neoliberal-driven development discourses in tourism culminate in a hyper-neoliberal spiral that "dominates the socio-technical system and frames the policy agenda that affect directly and indirectly impact tourism destinations and DMOs and sets destinations on particular development trajectories" (ibid., p. 12).

Studies from developing and developed countries show different actions and roles of government authorities at different scales. For example, findings from Namibia indicate that the national government plays an important role in supporting local communities to benefit from tourism activities (Kavita & Saarinen, 2016). Similarly, the federal regionalization tourism programme implemented in Brazil has contributed to fostering economic growth, particularly in urban areas (Silva et al., 2023). Local authorities can often have an important coordinating role in the promotion and development of tourism, which is the case in Tuscany, Italy (Gori et al., 2021). Research has also provided evidence on the role of governments in the promulgation of laws for land zoning and investment opportunity areas for tourism (Dela Santa, 2018; Gotham, 2012) and of legislation or municipal action to regulate tourism activities at the local scale. This has especially been the case when seeking to manage the impacts of Airbnb and Uber (Amore et al., 2022; Hall et al., 2022), as well as the setting of service standards in the hospitality sector (e.g., the Canary Islands, Spain) (Simancas Cruz et al., 2018). Yet, when it comes to environmental sustainability and tourism policies, the coordinating role of the state at different scales can be far from effective (Brendehaug et al., 2017) or limited to the will of powerful industry stakeholders (Farmaki et al., 2015). Direct taxation from tourism activities to favour de-carbonization has been suggested as a possible solution to reduce the environmental externalities of the sector (Scott et al., 2016), even though examples from highly congested destinations suggests that its effectiveness to address environmental issues is questionable (Nepal & Nepal, 2021).

The role of government and public authorities in tourism has also come under scrutiny because of the escalating climate crisis and the COVID-19 pandemic (Gössling et al., 2021; Kennell, 2020). The implementation of government-funded business aid and recovery schemes (e.g., furlough, monetary incentives) and the regulation of lockdowns to restrict national and international travel are testaments of the interventionist role and prominence of the state in regards to markets, mobility, resources, and societies, despite them seemingly being at odds with the market-driven policies that were implemented over the last decades (Kennell, 2020; Seyfi et al., 2023). However, Gössling et al. (2021, p. 3) warned on the "urge by many to go back to business as usual, perhaps to overcompensate for losses by even more aggressive growth" as the global COVID-19 pandemic crisis de-escalated. Amore and Adie (2021) acknowledged the "new normal" dangerously leading to an exacerbation of the same old social and economic disparities in destinations, particularly in urban areas. Although there were pleas for alternative policies towards a just pathway for tourism in respect of communities and ecological limits (Higgins-Desbiolles, 2020, 2021), there was "no evidence-based strategy for climate change mitigation, and an overall silence regarding pandemic and other risks the global tourism system imposes on itself and the global economy" (Gössling et al., 2021, p. 13). The academic arguments over the ideal balance between economic development and sustainability (Higgins-Desbiolles, 2020) did not effectively enter public policy arenas and mirrors the lack of progress in addressing the climate crisis in tourism policymaking.

Data Collection

For the purposes of this study, we collected policy documents and tourism strategies enacted by national governments following the COVID-19 outbreak and global health crisis starting in March 2020. The focus on documents and strategies by national authorities is not new to tourism policy analysis and research (Dredge & Jenkins, 2012; Torkington et al., 2020). Tourism strategies as a policy output can be regarded as episodes of governance (Healey, 2003, 2006) that are available to the scrutiny and critical appraisal of scholars devoid of covert forms of power and influence between the researcher and the researched. As Hall (2010b, p. 40) argues, "policy is more than just a written document, although that may represent an important output of a decision- and policy-making process".

The analysis focused on national tourism strategies issued by branches of governments responsible for tourism development in El Salvador, Italy, the Maldives, South Africa, and the United States (USA) (Table 15.1). The selection was based on geographical representation, travel, and tourism policies conditions (WEF, 2022) and share of the travel and tourism industry to the economy in 2019 (WTTC, 2023). Additionally, we ensured a homogenous representation of destination contexts with diverse travel restriction policies implemented during the COVID-19 pandemic (Mathieu et al., 2020). A further criterion for the selection was the tourism strategy implementation period, with a start date during or after the COVID-19 pandemic breakthrough. Tourism strategy extracts were coded with the support of NVivo and grouped under pre-established categories adapted from Hall (2008).

Findings

Tourism as a Policy Issue

The coordinating role of national governments is acknowledged as a factor in the implementation of the tourism strategy. This can be found in the tourism strategy for El Salvador, which mentions the coordinating role of the Tourism Cabinet (*Gabinete Turistico*) in supporting the Ministry of Tourism to implement the strategy. This is further emphasized in one of the pillars of the strategy (*Institucionalidad del Sector*) which seeks to strengthen the action of the government in terms of political and regulatory frameworks (Gobierno de El Salvador, 2020). The tourism strategy for Italy also recognizes the importance of the public sector as a coordinator and its role in ensuring greater cohesiveness and reducing governance clashes

Table 15.1 List of selected countries for data collection and analysis

Country	Region (WEF, 2022)	T&T Policies ranking (WEF, 2022)	%/GDP (2019) (WTTC, 2023)	Travel restriction policies implemented (Mathieu et al., 2020)	Strategy (Years)
El Salvador	The Americas	47th	11.7	Restricted (Mar–Sep 2020) Required (Sep 2020–Mar 2021)	2020–2030
Italy	Europe	55th	10.6	Recommended (Feb 2020) Required (Mar–Jun 2020) Recommended (Jul 2020) Recommended (Aug 2020–July 2021)	2023–2027
Maldives	Asia-Pacific	n/a	51.4	n/a	2023–2027
South Africa	Sub-Saharan Africa	41st	7.8	Recommended (Mar 2020) Required (Apr–Aug 2020) Recommended (Aug–Oct 2020)* Required (July 2021)	2020–2025
USA	The Americas	40th	9.0	Recommended (Mar 2020) Required (Mar 2020–Mar 2021) Recommended (Apr–May 2021) Required (May–July 2021) Recommended (July 2021–)*	2022–2027

Notes: *Since November 2020, no travel restrictions have been imposed to international travel to the country, except a recommended restriction movement in July 2021

between national and regional authorities. In contrast, the New Zealand strategy places much more emphasis on public-private partnership led by the New Zealand Tourism Industry Association. The territorial governance for tourism in Italy is, in theory, under juridical and legislative responsibility of the regions, but the national government has put in place a joint regional body (*Conferenza dei Presidenti delle Regioni e delle Province Autonome di Trento e di Bolzano*) to work in concert with the Minister of Tourism and the associations of provincial and municipal authorities. The latter are directly involved to work in coordination with the Minister to address local level issues like transport logistics and the hosting of MICE events and hallmark cultural events (Parlamento Italiano, 2023). A similar coordinating role can be found in the strategies for the Maldives and the United States, with an emphasis on the role of government ministries to "collaborate across and within the public sector to prioritise the tourism industry" (Republic of Maldives, 2023, p. 72) and to "promote sustainable tourism in coordination with regional, state, local, and tribal tourism and outdoor recreation offices" (Government of the United States of America, 2022, p. 22), while the New Zealand government has a more limited coordinating role overall. However, in New Zealand, as in other jurisdictions, "incentives" for coordination between central government, regional tourist organizations, and industry is encouraged through funding programmes for promotion and development.

When looking at the definition of *stakeholders*, it comes as no surprise that the strategies tend to refer almost exclusively to industry partners. For instance, the tourism plan for El Salvador mentions the private sector as the main stakeholder for government entities to collaborate with towards the development of tourism in the country (Gobierno de El Salvador, 2020). In the case of the Maldives, the 2023–2027 Tourism Master Plan was developed in consultation with industry stakeholders and partners (Republic of Maldives, 2023). The 2021–2025 strategic plan for tourism in South Africa stresses the pursuit of a corporatist governance approach encompassing strategic partnerships and collaborations with industry stakeholders (Republic of South Africa, 2020). The two notable exceptions come from the US tourist strategy, which explicitly mentions community stakeholders be consulted "to plan for the most desired improvements, identify cross-management or sustainability needs, and advocate for funding that supports capacity-building efforts" (Government of the United States of America, 2022, p. 21). The strategy also acknowledges the importance of involving tribal communities as a way to implement the vision of the strategy in respect of the *Native American Tourism and Improving Visitor Experience Act* (Government of the United States of America, 2022). Similarly, New Zealand specifically acknowledges the role of Māori and Māori tourism organizations as stakeholders.

The actions of state authorities in tourism across the observed countries shows a convergence in planning practice. In particular, there is a narrative that conveys destination plans as business strategies and provides a range of targets and goals to be achieved within a period between four (e.g., Italy) and ten years (El Salvador). The tendency in the analysed documents is to identify a range of main goals and targets, with a preliminary identification of what performance criteria to monitor and the relevant public sector stakeholder responsible for the monitoring and reporting. The tendency, however, is to define tourism development-specific goals and targets without a proper acknowledgement of wider macroeconomic policy agendas and issues. The only document that explicitly acknowledges the importance of framing tourism within the wider development of the country can be found in the Italian destination strategy, which recognizes how the 2023–2027 tourism plan should be adapted and implemented in line with the joint EU-Italy *Piano Nazionale di Ripresa e Resilienza* (PNRR). The latter also highlights the potential role of transnational organizations or institutional arrangements, such as the European Union, in influencing national, regional, and local tourism policies.

The impact of the COVID-19 pandemic on the tourism sector and the need for some form of direct interventions of state authorities in legislation, regulation, and taxation is addressed across the plans and strategies. In the case of Italy, a dedicated scheme is enacted to ensure tourism businesses can restructure their debts and thus avoid bursting following the years of high uncertainty and forced temporary business closure following the pandemic (Parlamento Italiano, 2023). Additional measures encompass the reform of the regional tour guides system, the standardization of the existing hotel classifications and certifications, and incentives to hire new employees in the sector. In the Maldives, the strategy plan foresees the allocation of a green tax revenue for businesses seeking to invest in green infrastructure as well as mechanisms to increase access to finance for SMEs and reduce red tape for registering a new property (Republic of Maldives, 2023). The national tourism strategy for the United States is the only one explicitly mentioning "climate change resilience and adaptation strategies into infrastructure investment and management and visitor use management on federal lands and waters" (Government of the United States of America, 2022, p. 32). Yet, the legislative extent of federal and state sphere of action and the regulatory guidelines that are required to effectively implement strategies are not addressed, with similar issues of implementation also occurring in the New Zealand context.

Almost all the strategies retrieved for this study recognize the importance of investing in infrastructure as one of the key priorities. Both the Italian and the South African tourism strategies refer to MICE tourism as a target for infrastructural and direct government support in bidding and hosting applications (Parlamento Italiano, 2023; Republic of South Africa, 2020). The Italian and the Maldivian strategies also acknowledge the necessity to enhance transport infrastructure investments for maritime (Parlamento Italiano, 2023) and airport transport (Republic of Maldives, 2023) respectively. Focusing on financial incentives, the tourism strategy for Italy defers to the National Recovery and Resilience Plan's (*Piano Nazionale di Ripresa e Resilienza* – PNRR) dedicated funding scheme for new projects aimed at enhancing digitalization, innovation, and environmental sustainability practices in tourism services (Parlamento Italiano, 2023). In comparison, the strategy for the Maldives is for the establishment of a dedicated fund to assist climate adaptation measures in the sector (Republic of Maldives, 2023). Conversely, the tourism strategy for South Africa identified the Tourism Incentive Programme as the already existing funding scheme to assist tourism activities (Republic of South Africa, 2020). In a similar fashion, the tourism strategy for the United States refers to existing federal funding schemes to be communicated and made available to small-scale tourism providers (Government of the United States of America, 2022). The New Zealand strategy also noted the importance of infrastructure, but a subsequent change of government since the strategy was developed also highlights the difficulties of effective strategy planning for tourism.

The rhetoric of tourism as a key contributor towards economic development permeates the strategies of the observed destinations. The strategies for New Zealand, South Africa, and El Salvador stressed the positive outcomes in terms of employment, visitor spending, increased tax revenues, and foreign trade and currency (Gobierno de El Salvador, 2020). Similarly, the strategy for the Maldives considers tourism as key for the prosperity of the country and identifies priority goals and objectives to strengthen the competitiveness of the destination on the global stage (Republic of Maldives, 2023). The tourism strategy for Italy reiterates the competitiveness dogma and how reforms in tourism should promote green development and circular economy solutions for tourism to be competitive (Parlamento Italiano, 2023). Conversely, socio-environmental development goals through tourism are only explicitly acknowledged in the strategies for El Salvador and the United States. On the one hand, the Government of El Salvador acknowledges the necessity to find a balance between economic development and the conservation of natural and cultural resources (Gobierno de El Salvador, 2020). On the other hand, the Government of the United States advocates for the implementation of travel and tourism projects featuring "environmental justice, inclusion, and resilience principles" (Government of the United States of America, 2022, p. 23).

Overall, the tourism strategies featured in this analysis present a positivistic appraisal of tourism-driven development and the "grow-ism" rhetoric that much characterized tourism policies prior to the COVID-19 outbreak and global pandemic. The latter is only acknowledged in terms of decreased tourism volume (e.g., Maldives, United States). Most strikingly, ecological vulnerabilities and socio-economic challenges attributed to tourism prior to the pandemic are overlooked. This is somehow bizarre if we consider that Italy is home to the number one overtourism city (Amore et al., 2020), the United States have been significantly exposed to natural hazards (e.g., hurricanes) (Adie, 2024), and the Maldives have long been coping with sea-level rising and coral bleaching as a consequence of climate change (Scheyvens, 2011). The question, therefore, is whether we are in a situation of a missed opportunity and a reiteration of a policy-as-usual scenario.

Directions

Given the hollowing out of the state and the subsequent focus on governance and coordination the question rightly applies as to how interventions can encourage destinations to take different trajectories as opposed to boosterism and economically driven and fast-track recovery approaches. These issues have become even more important given the interest in the process of sustainable transitions and the advocacy towards a much-needed transformative shift in current socio-ecological regimes. These processes are inherently political because as Hölscher et al. (2018, p. 2) note, "actors play key roles in shaping desirable transitions and transformations through transformative agency and governance. Processes to shape transitions and transformations are deeply political, involving power struggles and value conflicts". Furthermore, it is important to recognize that despite arguments for the transformative nature of external crisis events, such as COVID-19, from a policy process perspective, the "outcome" of post-crisis pathways may not be transformative at all in the sense of being more sustainable, but instead revert to system maintenance (business-as-usual approaches). Thus, in times of crisis and emergency, destinations may revert to old ways of "doing" and sometimes doing it in a stronger fashion. This is especially significant in the aftermath of large-scale disasters whereby state encouragement for capital investment has only been heightened by the opportunities provided by disasters (Hall et al., 2016) and the continued commitment of governments to place competition strategies (Malecki, 2004). Nevertheless, an interesting future avenue in policy studies is seeking to understand the way in which policy interventions are linked to pathway experimentation and, perhaps of increased concern given environmental crises, pathway lock-in, i.e., how policy decisions at one point in time, such as in the case of infrastructure investment, set development directions for subsequent years which can only be broken out of at great cost (Rifkin, 2019).

Discussion and Conclusions

The nexus between hyperneoliberalism, governance, and tourism policy sets the context for contemporary policy interventions. Such relationships are not always negative and have been highlighted in relation to the pursuit of sound good governance principles of equality, justice, and democracy among marginalized Mayan communities in Quintana Roo, Mexico (Jamal & Camargo, 2014). However, at the same time, it is crucial to prioritize ecological protection, rectify the inequalities caused by neoliberal-infused tourism policy, and put "tourism in the service to the public and to be accountable to the public" (Higgins-Desbiolles, 2020: 618). As Jamal and Camargo (2014: 26) put it:

> We lack theoretical and practical tools to tackle important ethical and justice-related issues related to destination development and marketing, particularly intangible aspects such as human–environmental (ecocultural) relationships, cultural commodification, and inequitable distribution of tourism opportunities (and costs) among disadvantaged groups … We call for an ecocultural, participatory, and integrated framework of justice and care to guide sustainable tourism.

The readers of this chapter might wonder how such third-order change can be achieved when policy actors are unwilling to acknowledge policy failure or recognize potential policy alternatives. Undeniably, this is a hefty task for tourism scholars to accomplish. Applied research in public policy occurs within a depoliticized environment of consensus, rationalization, and power (Cryle & Hillier, 2005; Flyvbjerg, 1998) in which policy options and pathways are presented as the result of "rational" policy processes, and tourism is no exception. Nonetheless, it is the task of scholars to critically conduct policy analysis and question the legitimacy and rationale of policy actions against a range of normative standards with respect to the environment, stakeholder recognition and participation, and ethics. The way forms of citizenry engagement and inclusive participation in tourism policy and governance are evolving (Erdmenger, 2022) also reflect obsolete modes of destination governance. At the same time, research should recognize the powerless and those seeking to have their voices heard, in the pursuit of genuinely democratic, equal, and diverse governance arenas to improve policy outcomes and impacts.

References

Aall, C., Dodds, R., Sælensminde, I., & Brendehaug, E. (2015). Introducing the concept of environmental policy integration into the discourse on sustainable tourism: A way to improve policy-making and implementation? *Journal of Sustainable Tourism*, *23*(7), 977–989.

Adie, B. A. (2019). *World Heritage and Tourism: Marketing and Management*. Routledge.

Adie, B. A. (2024). Fighting mother nature: Second home owners, risk awareness, and post-disaster planning on Fire Island, New York. In B. A. Adie & C. M. Hall (Eds.), *Second Homes and Climate Change* (pp. 73–86). Routledge.

Amore, A. (2023). The importance of policies and planning in the tourism job market: A reflection on actions and inactions in the United Kingdom. In P. Xie (Ed.), *Handbook on Tourism Planning* (pp. 182–193). Edward Elgar.

Amore, A. (forthcoming). Tourism, public policy and governance. In C. M. Hall (Ed.), *The Wiley Blackwell Companion to Tourism*. Wiley.

Amore, A., & Adie, B. A. (2021). (Re)igniting tourism in cities after COVID: The same old risks of the "new normal". In V. Pecorelli (Ed.), *From Overtourism To Undertourism: Any Sustainable Scenarios in the Post Pandemic Time?* (pp. 13–31). Unicopli.

Amore, A., de Bernardi, C., & Arvanitis, P. (2022). The impacts of Airbnb in Athens, Lisbon and Milan: A rent gap theory perspective. *Current Issues in Tourism, 25*(20), 3329–3342.

Amore, A., Falk, M., & Adie, B. A. (2020). One visitor too many: Assessing the degree of overtourism in established European urban destinations. *International Journal of Tourism Cities, 6*(1), 117–137.

Amore, A., & Hall, C. M. (2017). National and urban public policy agenda in tourism: Towards the emergence of a hyperneoliberal script? *International Journal of Tourism Policy, 7*(1), 4–22.

Amore, A., Hall, C. M., & Jenkins, J. M. (2017). They never said "Come here and let's talk about it": Exclusion and non-decision-making in the rebuild of Christchurch, New Zealand. *Local Economy, 32*(7), 617–639.

Bohlin, M., Brandt, D., & Elbe, J. (2016). Tourism as a vehicle for regional development in peripheral areas – myth or reality? A longitudinal case study of Swedish regions. *European Planning Studies, 24*(10), 1788–1805.

Bramwell, B. (2011). Governance, the state and sustainable tourism: A political economy approach. *Journal of Sustainable Tourism, 19*(4–5), 459–477.

Bramwell, B., & Meyer, D. (2007). Power and tourism policy relations in transition. *Annals of Tourism Research, 34*(3), 766–788.

Brendehaug, E., Aall, C., & Dodds, R. (2017). Environmental policy integration as a strategy for sustainable tourism planning: Issues in implementation. *Journal of Sustainable Tourism, 25*(9), 1257–1274.

Brouder, P. (2020). Reset redux: possible evolutionary pathways towards the transformation of tourism in a COVID-19 world. *Tourism Geographies, 22*(3), 484–490.

Cryle, D., & Hillier, J. (Eds.) (2005). *Consent and Consensus. Politics, Media and Governance in Twentieth Century Australia*. Australia Research Institute.

Dela Santa, E. (2018). Fiscal incentives for tourism development in the Philippines: A case study from policy networks and advocacy coalition framework. *Tourism Planning & Development, 15*(6), 615–632.

Dodds, R., & Butler, R. W. (2008). Inaction more than action. In S. Gössling, C. M. Hall, & D. B. Weaver (Eds.), *Sustainable Tourism Futures: Perspectives on Systems, Restructuring and Innovations* (pp. 43–57). Routledge.

Dredge, D., & Jenkins, J. (2012). Australian national tourism policy: Influences of reflexive and political modernisation. *Tourism Planning & Development, 9*(3), 231–251.

Duffy, R. (2015). Nature-based tourism and neoliberalism: Concealing contradictions. *Tourism Geographies, 17*(4), 529–543.

Erdmenger, E. C. (2022). The end of participatory destination governance as we thought to know it. *Tourism Geographies, 25*(4), 1–23.

Farmaki, A., Altinay, L., Botterill, D., & Hilke, S. (2015). Politics and sustainable tourism: The case of Cyprus. *Tourism Management, 47*, 178–190.

Flyvbjerg, B. (1998). *Rationality and Power: Democracy in Practice*. Chicago University Press.

Gobierno de El Salvador (2020). *Plan Nacional de Turismo El Salvador 2030*. Gobierno de El Salvador.

Gori, E., Fissi, S., & Romolini, A. (2021). A collaborative approach in tourism planning: The case of Tuscany region. *European Journal of Tourism Research, 29*, 2907.

Gössling, S., Scott, D., & Hall, C. M. (2021). Pandemics, tourism and global change: A rapid assessment of COVID-19. *Journal of Sustainable Tourism, 29*(1), 1–20.

Gotham, K. F. (2012). Disaster, Inc.: Privatization and post-Katrina rebuilding in New Orleans. *Perspectives on Politics, 10*(3), 633–646.

Government of the United States of America. (2022). *National Travel Tourism Strategy*. Government of the United States of America.

Hall, C. M. (1999). Rethinking collaboration and partnership: A public policy perspective. *Journal of Sustainable Tourism, 7*(3–4), 274–289.

Hall, C. M. (2008). *Tourism Planning: Policies, Processes and Relationships*. Pearson.

Hall, C. M. (2010a). Politics and tourism – interdependency and implications in understanding change. In R. Butler & W. Suntikul (Eds.), *Tourism and Political Change* (pp. 7–18). Goodfellows Publishing.

Hall, C. M. (2010b). Researching the political in tourism. In C. M. Hall (Ed.), *Fieldwork in Tourism: Methods, Issues and Reflections* (pp. 39–54). Routledge.

Hall, C. M. (2019). Constructing sustainable tourism development: The 2030 agenda and the managerial ecology of sustainable tourism. *Journal of Sustainable Tourism, 27*(7), 1044–1060.

Hall, C. M., & Amore, A. (2016). Turismo, sostenibilità e crescita verde: Green Economy o una semplice pennellata di verde? In A. Pecoraro Scanio (Ed.), *Turismo Sostenibile. Retorica e Prartiche* (pp. 145–188). Aracne Editrice.

Hall, C. M., & Amore, A. (2020). The 2015 Cricket World Cup in Christchurch: Using an event for post-disaster reimagine and regeneration. *Journal of Place Management and Development, 13*(1), 4–17.

Hall, C. M., & Jenkins, J. M. (1995). *Tourism and Public Policy*. Routledge.

Hall, C. M., Malinen, S., Vosslamber, R., & Wordsworth, R. (Eds.). (2016). *Business and Post-disaster Management: Business, Organisational and Consumer Resilience and the Christchurch Earthquakes*. Routledge.

Hall, C. M., Prayag, G., Safonov, A., Coles, T., Gössling, S., & Naderi Koupaei, S. (2022). Airbnb and the sharing economy. *Current Issues in Tourism, 25*(19), 3057–3067.

Hall, C. M., Scott, D., & Gössling, S. (2020). Pandemics, transformations and tourism: Be careful what you wish for. *Tourism Geographies, 22*(3), 577–598.

Hall, C. M., & Seyfi, S. (2021). Covid-19 pandemic, tourism and degrowth. In C. M. Hall, L. Lundmark, & J. J. Zhang (Eds.), *Degrowth and Tourism: New Perspectives on Tourism Entrepreneurships, Destinations and Policy* (pp. 220–239). Routledge.

Healey, P. (2003). Collaborative planning in perspective. *Planning Theory, 2*(2), 101–123.

Healey, P. (2006). *Urban Complexity and Spatial Strategies: Towards a Relational Planning for Our Times*. Routledge.

Higgins-Desbiolles, F. (2011). Death by a thousand cuts: Governance and environmental trade-offs in ecotourism development at Kangaroo Island, South Australia. *Journal of Sustainable Tourism, 19*(4–5), 553–570.

Higgins-Desbiolles, F. (2020). Socialising tourism for social and ecological justice after COVID-19. *Tourism Geographies, 22*(3), 610–623.

Higgins-Desbiolles, F. (2021). The "war over tourism": Challenges to sustainable tourism in the tourism academy after COVID-19. *Journal of Sustainable Tourism, 29*(4), 551–569.

Hölscher, K., Wittmayer, J. M., & Loorbach, D. (2018). Transition versus transformation: What's the difference? *Environmental Innovation and Societal Transitions, 27*, 1–3.

Jamal, T., & Camargo, B. A. (2014). Sustainable tourism, justice, and an ethic of care: Toward the just destination. *Journal of Sustainable Tourism*, *22*(1), 11–30.

Kavita, E., & Saarinen, J. (2016). Tourism and rural community development in Namibia: Policy issues review. *Fennia - International Journal of Geography*, *194*(1), 79–88.

Kennell, J. (2020). Tourism policy research after the COVID-19 pandemic: Reconsidering the role of the state in tourism. *Skyline Business Journal*, *16*(1), 68–72.

Malecki, E. J. (2004). Jockeying for position: What it means and why it matters to regional development policy when places compete. *Regional Studies*, *38*(9), 1101–1120.

Mathieu, E., Ritchie, H., Rodés-Guirao, L., Appel, C., Giattino, C., Hasell, J., Macdonald, B., Dattani, S., Beltekian, D., Ortiz-Ospina, E., & Roser, M. (2020). Coronavirus pandemic (COVID-19). *Our World in Data*. https://ourworldindata.org/coronavirus.

Nepal, R., & Nepal, S. K. (2021). Managing overtourism through economic taxation: Policy lessons from five countries. *Tourism Geographies*, *23*(5–6), 1094–1115.

OECD (2022). *OECD Tourism Trends and Policies 2022*. OECD.

Parlamento Italiano (2023). *Schema del Piano Strategico di Sviluppo del Turismo per il Periodo 2023–2027*. Parlamento Italiano.

Republic of Maldives (2023). *Maldives Fifth Tourism Master Plan 2023–2027. Goals and Strategies*. Republic of Maldives.

Republic of South Africa (2020). *Stategic Plan 2020/21–2024/25*. Pretoria, Department of Tourism.

Rifkin, J. (2019). *The Green New Deal: Why the Fossil Fuel Civilization will Collapse by 2028, and the Bold Economic Plan to Save Life on Earth*. St. Martin's Press.

Rogerson, C. M. (2015). Tourism and regional development: The case of South Africa's distressed areas. *Development Southern Africa*, *32*(3), 277–291.

Scheyvens, R. (2011). The challenge of sustainable tourism development in the Maldives: Understanding the social and political dimensions of sustainability. *Asia Pacific Viewpoint*, *52*(2), 148–164.

Scott, D., Gössling, S., Hall, C. M., & Peeters, P. (2016). Can tourism be part of the decarbonized global economy? The costs and risks of alternate carbon reduction policy pathways. *Journal of Sustainable Tourism*, *24*(1), 52–72.

Seyfi, S., Hall, C. M., & Shabani, B. (2023). COVID-19 and international travel restrictions: The geopolitics of health and tourism. *Tourism Geographies*, *25*(1), 357–373.

Silva, T. C., da Silva Neto, P. V., & Tabak, B. M. (2023). Tourism and the economy: Evidence from Brazil. *Current Issues in Tourism*, *26*(6), 851–862.

Simancas Cruz, M., García Cruz, J. I., Greifemberg, C. A., & Peñarrubia Zaragoza, M. P. (2018). Strategies to improve the quality and competitiveness of coastal tourism areas. *Journal of Tourism Analysis*, *25*(1), 68–90.

Simmons, R. H., Davis, B. W., & Sager, D. D. (1974). Policy flow analysis: A conceptual model for comparative public policy research. *The Western Political Quarterly*, *27*(3), 457–468.

Torkington, K., Stanford, D., & Guiver, J. (2020). Discourse(s) of growth and sustainability in national tourism policy documents. *Journal of Sustainable Tourism*, *28*(7), 1041–1062.

UNWTO (2023). *Investing in People, Planet and Prosperity*. UNWTO.

World Economic Forum (WEF) (2022). *Travel & Tourism Development Index 2021: Rebuilding for a Sustainable and Resilient Future*. WEF.

World Travel and Tourism Council (WTTC) (2023). *Economic Impact Research*. WTTC. https://wttc.org/research/economic-impact.

Zahra, A. L. (2011). Rethinking regional tourism governance: The principle of subsidiarity. *Journal of Sustainable Tourism*, *19*(4–5), 535–552.

16 Two Faces of the Adriatic Pearl

The Leader in Overtourism and Sustainability

Tina Šegota

Introduction

It is indisputable that the presence of tourists changes places – for the better, most would claim. However, it was not until recently that development through tourism was approached with a certain level of caution, which was formed due to epochal consecutive events: overtourism and the COVID-19 pandemic. In the decade preceding 2020, multiple places ranging from natural protected areas to urban hotspots were reported to be overwhelmed by tourists (Liberatore et al., 2022; Milano, 2017). A combination of 'Instagrammable' beauty (Siegel et al., 2023a, 2023b) and tourism governance seizing profitable opportunities from tourists' desires for travel and experiences (Šegota, 2018a) led to residents protesting against authorities, tourism development, and tourists (i.e., an antitourism movement began) (Martín et al., 2018). To mitigate this antitourist movement, governments introduced various measures to tackle the influx of tourists and restore residents' satisfaction.

However, in March 2020, all this dramatically stopped when humanity faced the fight against the COVID-19 virus. While phrases such as 'social distancing', 'stay at home', 'avoid crowds', and 'physical distance' have dominated worldwide media headlines, various governments imposed restrictive measures preventing many human-to-human contact situations (Arslan & Allen, 2022; Huang et al., 2020). Hence, many events have been cancelled or postponed, just like other leisure and travel activities (Karagöz et al., 2023; Li et al., 2021). At this point, overtourism quickly became no tourism for many places, having significant negative socioeconomic consequences.

One such place is the city of Dubrovnik. The town has been paraded as the Pearl of the Adriatic since 1979 when it was endowed with the UNESCO Heritage Site title. The chapter will initially touch upon the importance of culture and heritage as the main drivers of Dubrovnik's tourism development. Then, the focus of the chapter will shift towards cruising and film-induced tourism – i.e., the two types of tourism that dominated the pre-pandemic decades of the city's tourism development. The focus of the discussion will be (mal)practices in tourism development and promotion on the local and national levels that led towards the city becoming a synonym for overtourism. We will present the 'Respect the City' initiative, which represents a 'multidisciplinary strategic destination management project with an

DOI: 10.4324/9781003449027-19

action plan. It is comprised of a set of short-term, medium-term and long-term measures and activities focused on Dubrovnik as the leader in sustainable and responsible tourism in the Mediterranean' (Dubrovnik Tourist Board, n.d.). The Respect the City strategic activities implemented pre-pandemic will be critically evaluated, in addition to discussing local and national tourism development (mal) practices post-pandemic. For example, while Respect the City focuses on limiting one-day visitors from cruise ships to the city, its tourist board is promoting the city being crowned the Best European Cruise Destination 2021 by the renowned World Travel Awards. The latter feels as if Dubrovnik initially bit the hand that fed it before the pandemic and very soon bit its tongue when it was hit the hardest among Croatian tourist destinations during the pandemic. Hence, the critical evaluation of Respect the City and post-pandemic activities will focus on whether the city has been rightfully self-proclaimed as the leader of sustainable and responsible tourism development in the Mediterranean.

Dubrovnik and its Identity as the Pearl of the Adriatic

A place in the southernmost region of Croatia has been receiving notable admiration for decades for its natural and cultural beauty. The city of Dubrovnik appears as a medieval town stretched along the narrow Adriatic coastal belt under the Dinaric mountain peaks. The city was formally established in the 12th century by the joining and fortifying of two Slavic settlements. This is the period that marked the start of building the City Walls (Visit Dubrovnik, n.d.a). The City Walls have been strengthened and fortified since, keeping Dubrovnik's residents and visitors safe from enemies, e.g., the Venetians, Ottomans, and even those from across the Adriatic Sea. Over the centuries, Dubrovnik became a significant Mediterranean Sea power, as evidenced by its many diverse buildings, such as Gothic, Renaissance, and Baroque churches, monasteries, palaces, and fountains (UNESCO, n.d.). In 1667, the city was severely damaged by an earthquake; although most of the Old Town has been preserved to the present day. However, the earthquake represents a shift in the Old Town's architecture which is still visible: there are not many balconies in the Old Town, and the remaining act as a reminder of how the city looked before the earthquake.

While the city of Dubrovnik has always been a rival to Venice, it joined the UNESCO List of World Heritage Sites in 1979, almost a decade earlier than its northern counterpart. Ever since joining the List, Dubrovnik has become a well-known tourist destination. The mountainous landscape, the crystal-clear blue sea, and the unique architecture have made Dubrovnik worthy of the nickname 'Pearl of the Adriatic'. What attracts tourists the most is the view of its preserved white stone defensive walls, endowed with numerous forts and towers, surrounding the medieval red-roof-topped houses and palaces (see Figure 16.1).

Dubrovnik's inclusion in the UNESCO List adds value to its destination branding. Being named a UNESCO World Heritage Site brings particular attention to a place's natural or cultural endowments (Cuccia et al., 2017). Such inclusion emphasises that the city holds an outstanding value of pure and original heritage

Figure 16.1 Wall, rooftops, and surroundings

worth preserving and admiring (Frey & Steiner, 2013). Moreover, it also signals that there is a specific unique experience worthy of the tourist gaze (Urry & Larsen, 2011) that influences the interpretation of authenticity. When people visit these places, they report that their lives have been enriched by the opportunity to see these unique destinations (Lisle, 2016; Urry & Larsen, 2011). The characteristics of these 'special' places create the atmosphere of the tourist experience, with effective change and transformation outcomes. Lowenthal (1985) says that localities with old buildings display solidity, continuity, authority, and craft; they have survived developers, town planners, wars, erosion, and earthquakes for years and years; they link past generations to the present; they demonstrate that tradition and age are worthy of preservation; they were built without the help of modern technologies. Once this is understood in the context of cultural heritage – something unique, irreplaceable, and unrepeatable – it is easy to understand why tourists visit places and gaze upon the objects, landscapes, and traditions in search of authentic and memorable experiences.

The inclusion on the World Heritage List is also a claim to authenticity, distinguishing the place from other heritage sites. Frey and Steiner (2013) claim that the designation of 'World Heritage' transforms places into major attractions and icons of national identity, and, as such, they become very popular. Destination management organisations (hereinafter DMOs) also use the World Heritage label to differentiate their specific tourism destinations from their competitors, presenting the sites as must-sees (de Fauconberg et al., 2017; Marcotte & Bourdeau, 2012). On the one hand, DMOs recognised that contemporary fascination with gazing upon extraordinary heritage contributes to economic growth (Arezki et al., 2009), but it also helps increase the attractiveness of the wider region, thus spreading and increasing the economic contribution (Frey & Steiner, 2013; Mazanec et al., 2007; Patuelli et al., 2016).

In the decade leading to the its inclusion on the World Heritage List, Dubrovnik was investing in tourism infrastructure and superstructure. In 1965, it reached over 1 million overnight stays (Kobašić, 1993), a significant tourism achievement. Two decades later, the whole Dubrovnik region recorded 5,868,240 overnight stays, of which 52.3% were international tourists (Kobašić, 1993). Kobašić (1993, p. 111) reports that the visitation boom was significant across the board: from 8.3% of tourism's contribution to Dubrovnik region's economy in 1960, there was an increased to 19.7% in 1970, 28.9% in 1980, and 33.9% in 1990. Also, the percentage of tourism and hospitality employees increased from 20.8% in 1968 to 35.9% in 1990.

In the 1990s, the city was yet again damaged during armed conflict. Naturally, the conflict also stopped its tourism development. It took Dubrovnik years to restore its tourism industry and to repair damages inflicted by the Homeland War. In restoring its Old Town, UNESCO played a significant role: it coordinated the restoration so that the unique architecture was preserved, and the city held onto its Heritage Site status. In the middle of the 1990s, the nowadays distinctive red rooftops emerged. Dubrovnik's roofs were initially red, but their colour faded over centuries. However, it was not until after the devastating damages of the war and the restoration that they were reinstated to bright red as we see them today (see Figure 16.2). The repair has played a significant role in leading to Dubrovnik being proclaimed as the King's Landing, a fictional kingdom's capital, which sealed the the city's worldwide fame.

Cruise- and Film-Induced Tourism – Recovery Leading to a Deathtrap

The Perks of Cruise- and Film-Induced Tourism

The period after the Homeland War has brought two significant changes to Dubrovnik. Firstly, the restoration under UNESCO took its course, coupled with the general repair of the place, the revival of its economy in general, and the tourism and hospitality industry in particular, with many international tourists again visiting Dubrovnik and some of its residents returning from exile. Dubrovnik started redeveloping and reviving its tourism industry at the beginning of the 2000s, and it only reached the pre-war tourist numbers in 2009. It was not until 2014 when, for the first time, the number of overnight stays was higher than the record-breaking number of 2,691,726 overnight stays recorded in 1987 just for the city of Dubrovnik (Ban & Vrtiprah, 1999; Vrtiprah & Dragičević, 2017). However, 20 years of difference also changed the length of visits. For example, tourists visiting the city of Dubrovnik stayed in the town and its surroundings on average for 5.2 days, while in 2014, the visitation was fast-paced and shorter, with an average

Figure 16.2 Distinctive red rooftops

of 3.4 days (Vrtiprah & Dragičević, 2017). Dubrovnik also saw the opportunity to develop into a cruising destination. Just like its northern Adriatic historical rival Venice, Dubrovnik started welcoming cruise ships, and with a year-on-year growth of 11.5%, the total number of cruise ship passengers visiting exceeded 1 million in 2010 (Vrtiprah & Dragičević, 2017). Most tourists arrived by bigger ships that could hold more than 3,000 passengers, resulting in over 1,176,086 arrivals on 688 boats in 2013 (Vrtiprah & Dragičević, 2017).

Besides the tourism redevelopment related to cruise ships, Dubrovnik also saw the opportunity in film tourism, which was the second significant change that shaped the city after the war. In 2011, the filming of the highly successful TV series *Game of Thrones* started in Dubrovnik. The original series was based on George R. R. Martin's bestselling fantasy novels, which portray chronicles of violent dynastic struggle among noble families while more threats emerge from the north and east (Šegota, 2018a, 2018b). With a production budget of up to US$100 million per season and 38 Emmy awards, this record-setting TV series commands an average of 20 million viewers per episode, making it the most successful TV series in history (HBO, 2016). HBO producers decided to relocate some filming locations from Malta to Croatia, especially those related to the King's Landing, the most crucial place in the series, known for its Iron Throne. Most of the filming occurred in the Old Town of Dubrovnik and at nearby tourist attractions because of the distinctive red rooftops, walls, and towers.

Croatian DMOs instantly recognised the economic benefits of tourism driven by a desire for an authentic experience of on-screen location sites in Dubrovnik and its surroundings. Hence, Croatian DMOs strongly emphasised links between Dubrovnik and the fictional King's Landing in online and offline destination marketing activities. As a result, Dubrovnik became the pilgrimage destination for fans of *Game of Thrones* (Šegota, 2018a, 2018b).

In a period dominated by cruise and film tourism, Dubrovnik boasted increasing tourist visits. For example, in 2015, there were almost one and a half million tourist arrivals, more than 860,000 cruise passengers on 600 boats, nearly 3 million overnight stays, and nearly 1 million admission tickets sold for the City Walls (Šegota, 2018b; Tkalec et al., 2017; Vrtiprah & Dragičević, 2017). Tkalec et al. (2017) attribute these record tourist numbers predominately to HBO's 'megahit' TV series, suggesting that *Game of Thrones* film-induced tourism resulted in a 37.9% increase in tourist arrivals, a 28.5% increase in overnight stays, and a 37.5% increase in City Walls admission tickets purchased, compared to a mere 7% increase in tourist arrivals, an 8.2% increase in overnight stays, and a 2.1% increase in City Walls admission tickets purchased in the pre-*Game of Thrones* period.

Tkalec et al.'s (2017) projections are theoretically very significant. The research on film-induced tourism has shown that localities featured in popular culture play an essential role in attracting visitors. From visitors' perspectives, when such sites are visited, people are able to gaze at the scene to relive the elements or certain aspects of the events conveyed through popular culture media. From the destination marketing perspective, the benefits of promoting localities through popular culture media are unparalleled (Beeton, 2010; Larson et al., 2013; Li et al., 2016; Šegota, 2018a). Bolan and Williams (2008) argue that the most significant benefit is reaching an audience via film imagery that would typically be unreachable through traditional marketing activities. First, a destination's exposure through film generates awareness of a destination among viewers whom other tourism marketing tools may not address. If a film is commercially successful, the market reach is even greater. Just like in the example of *Game of Thrones*, Croatia reached an audience of 10 million viewers per episode, with a total of 73 episodes. Such exposure is unimaginable to achieve in a decade for Dubrovnik and Croatian DMOs in terms of costs and reach.

Secondly, popular culture is considered an autonomous destination image formation agent, providing substantial information about a destination within a limited time. Since popular culture is often regarded as independent of the direct influence of DMOs, the information provided through the film is likely to be evaluated as more objective and unbiased (Šegota, 2018a). Lastly, film imagery, which includes on-screen virtual characters, an appealing storyline, memorable music, and remarkable landscapes, generates out-of-the-ordinary experiences which are then materialised in the moment a tourist visits the destination and recreates similar experiences as those viewed in on-screen performances (Bolan & Williams, 2008). Šegota (2018b, p. 117) asserts that 'the influence of film in emphasising the extraordinariness of "ordinary" must-see tourism products (that is, landscapes, places, and sites) is undeniable'.

The Perils of Cruise and Film-Induced Tourism

Being proclaimed as one of the most attractive Adriatic cruise destinations, it is unavoidable that the arrival and stay of cruising passengers affects the quality of life and the quality of the visitor experience in Dubrovnik. Even though it is one of the most important contributors to its growing economy, cruise tourism poses substantial challenges. Firstly, cruise visitors are not necessarily tourists, i.e., they do not stay overnight in the city (Bresson & Logossah, 2011). Hence, they become one-day visitors that, on the one hand, directly and indirectly financially benefit the local community but also contribute to crowding, traffic congestion, noise, waste, and irresponsible tourist behaviour (Del Chiappa et al., 2018; Jordan, 2014; Sindik et al., 2017). In the case of Dubrovnik, more than two-thirds of passengers from a single cruise ship visited the town's historic area, which often meant 9,000 cruise visitors at the tourist sites daily. If this volume is to be added to locally accommodated tourists and other one-day visitors to the Old Town, there is a significant breach of Dubrovnik's carrying capacity of 8,000 daily visitors (Puljić et al., 2019). This has created immense pressure for the city's infrastructure and resulted in overcrowding, significantly affecting residents' quality of life and visitor experience. As a result, the Old Town, once the most popular and prestigious location to live in, has lost a quarter of its local population (i.e., more than a thousand residents). Another challenge of Dubrovnik's cruise tourism is the uneven daily and yearly scheduling of ship calls. Most dockings in Dubrovnik's harbour take place between June to September, mainly for a single day or a few hours (Sindik et al., 2017; Vrtiprah & Dragičević, 2017), with some overlaps. This burdens the harbour and creates extreme traffic jams inwards and outwards of Dubrovnik.

Finding fame by being associated with one of the most successful TV shows in the history of the film industry is equally challenging and rewarding. The city has seen a more than 30% increase in tourism due to Dubrovnik's association with *Game of Thrones* (Tkalec et al., 2017), yet these record numbers create challenges for visitor management. Firstly, most visits occur in the summer, from June to September, and they usually include seeing the Old Town, the City Walls, and numerous fortresses that are all concentrated in a small geographical area. Moreover, just like cruise tourists, film-induced tourists may also exhibit a behaviour that evolves around fast-paced visits to filming sites and on-screen locations, adding to crowding, traffic congestion, noise, waste, and inappropriate tourist behaviour (Larson et al., 2013; Šegota, 2018b; Siegel et al., 2023b) (see Figure 16.3).

Hence, Dubrovnik's popularity had been proven too hard to handle. Partially induced by cruises and popular culture, Dubrovnik's tourism grew to unimaginable proportions in the period between 2013 and 2019 – it left the host community of 42,000 residents bestowed with the challenge of managing 1.1 million cruise passengers, 1.2 million international tourists, more than 4 million overnight stays in 50 hotels and 3,000 private apartments, and more than 10,000 hospitality employees (Puljić et al., 2019). Researchers observe that all this takes place in the summer months – from June to September – testifying to the destination's high tourism seasonality and Dubrovnik becoming a synonym for overtourism (Mandić et al., 2023; Puljić et al., 2019; Vrtiprah & Dragičević, 2017).

Figure 16.3 Tourists stage a photo shoot at the Wall

Respect the City – An Initiative for Mitigating Overtourism

Dubrovnik found itself crowned the most famous Croatian tourist destination. For some visitors, it still held the title Pearl of the Adriatic due to its unique cultural heritage, for which it was included in the UNESCO Heritage List. For some, it represented a fantasy land linked to their favourite TV show, and it became a film-induced pilgrimage site for *Game of Thrones* fans longing to visit a town that resembled King's Landing (Šegota, 2018b). For others, it was just a stop on their journey, worth the visit (Currie et al., 2008), regardless of whether they reached it by plane, cruise ship, or car. With tourists flooding the city and most of the tourist flows being concentrated around its main attractions, predominately found in and surrounding the Old Town, Dubrovnik started to face two major problems: loss of residents in the Old Town and its nearby area, and dissatisfaction with the quality of the visit from tourists. The loss of residents also meant houses and apartments were transformed into private tourist accommodation facilities, a global sharing economy trend (Paulauskaite et al., 2017) that did not bypass Dubrovnik.

Hence, the city officials decided to take active measures to mitigate the negative consequences of Dubrovnik's tourism popularity. Different and timely measures were united under the Respect the City (hereinafter RTC) initiative, which represents a 'multidisciplinary strategic destination management project with an action plan. It is comprised of a set of short-term, medium-term and long-term measures and activities focused on Dubrovnik as the leader in sustainable and responsible tourism in the Mediterranean' (Dubrovnik Tourist Board, n.d.). Puljić et al. (2019) observe that the initiative was first introduced in 2017 with the following goals:

1. Sustainable tourism development as a sector;
2. Sustainable use of the resources; and
3. Sustainable development for the people, the economy, and the community.

The initiative aimed at developing an integrated destination management plan for Dubrovnik to build capacity for private-public partnerships; foster communication, coordination, and cooperation among stakeholders; develop innovative solutions through optimisation, distribution, and diversification; and introduce and improve short- and long-term monitoring systems. Sustainability theory in tourism says that integrated destination management is achieved if all stakeholders partake in the activities, clearly and transparently communicating their agendas but also finding compromises for the better, longer, and more fruitful future of the community (Mihalič, 2020; Mihalič et al., 2016).

The RTC project envisioned that the stakeholders should be able to present themselves and work together through appreciation and alignment of ideas, perspectives, needs, and interests. Hence, the project encouraged the formation of the so-called working bodies that include advisory councils, working groups, and project consortia, aiming to develop solutions, innovations, products, and services through cooperation and co-investment of all stakeholders (Puljić et al., 2019).

The core strategies of the working bodies were the following:

• Promote the dispersal of visitors within the city and beyond;
• Stimulate new itineraries and attractions;
• Review and adapt regulations;
• Improve city infrastructure and facilities;
• Set monitoring and response measures.

Of course, not all stakeholders would be engaged in all aspects of the RTC; however, partnerships were expected to form between those most affected by specific strategic objectives. For example, to spread visitors within the city and across the region, four stakeholders were expected to partake in the action: the Dubrovnik Tourist Board, the Municipality of Dubrovnik, the Municipality of Konavle, and the Municipality of Dubrovačko Primorje. They were expected to work together in promoting attractions that would disperse visitors across three municipalities, encouraging visits to attractions in the neighbouring municipalities to Dubrovnik so as to avoid crowding in the Old Town. As a result, The Dubrovnik Tourist Board upgraded the Dubrovnik Card with three new attractions, all situated in the nearby town of Cavtat (Municipality of Konavle), with free transportation for the card holders.

Another example of strategically aligning regulations for better quality of life and avoidance of traffic congestion in the Old Town saw local police and utility companies (Libertas Dubrovnik and Sanitat Dubrovnik) being asked to improve coordination of arrivals and departures of buses, taxis, and delivery vehicles and thus implement new traffic regulations. The local police also worked with the Municipality of Dubrovnik to reduce crowding and regulate tourist activities in parts of the historic core by installing cameras and counting machines to control the entrance to the Old Town (see Dubrovnik Visitor, n.d.). This activity was critical, as exceeding the Old Town's carrying capacity of 6,000 visitors at any given moment meant that Dubrovnik's status as a UNESCO Heritage Site was jeopardised (Puljić et al., 2019).

One of the crucial strategic objectives was to minimise the negative impacts of cruise tourism. To limit one-day visitors from cruise ships to the city, and hence to address the carrying capacity concerns raised by UNESCO, the Municipality of Dubrovnik worked closely with Cruise Lines International Association and Dubrovnik Port Authority to ameliorate (dis)embankment time distribution and thus improve cruise ship arrival layovers. Moreover, or even more significantly, in 2019, cruise ships with up to 4,000 passengers were allowed to embank in Dubrovnik's Port Gruž, with an additional limit of up to two such cruises per day (Plan Bleu, 2022).

The initial RTC strategies and aligning objectives are presented in the UNWTO Overtourism Case Studies. Ever since, Dubrovnik's stakeholders have been working proactively, progressively, and cooperatively to ensure the conservation of cultural heritage, improvement of the quality of citizens' daily lives, and provision of the best possible experience of Dubrovnik as a destination for its visitors (Dubrovnik Tourist Board, n.d.).

The COVID-19 Pandemic: Consequences and Opportunities for Sustainable Tourism

The COVID-19 pandemic has had a devastating effect on people's health and economic development in general and completely gripped some service industries reliant on human resources like tourism (Karagöz et al., 2023; Li et al., 2021; Sharma et al., 2021). In the few years leading up to the pandemic, Dubrovnik developed and implemented some medium-term and long-term measures and activities, which were abruptly stopped in 2020. Just like other tourist places, and urban destinations in particular, Dubrovnik suffered significant loss of tourists. In the pre-pandemic record-breaking year of 2019, there were 1,415,006 tourist arrivals in the city of Dubrovnik, while the Old Town welcomed 997,348 visitors (Stojčić et al., 2023). In the summers of 2020 and 2021, when Croatia opened its borders and eased its COVID-19 restrictions and measures, Dubrovnik saw only 30–50% of its pre-pandemic tourist numbers (see Figure 16.4). The reason for such low numbers is that Dubrovnik is predominately a fly-in destination, with more than 50% of tourists arriving by plane (Vrtiprah & Dragičević, 2017). As flying was one of the first things banned or restricted during the pandemic, visitations to Dubrovnik dropped significantly, with results felt across its tourism and hospitality industry. In 2022, the number of tourist arrivals and overnight stays started to improve, reaching 75% of arrivals and 86% of overnight stays from the record-breaking 2019.

However, the pandemic did not have as devastating an effect on Dubrovnik's tourism as the Homeland War. During the war, no visits were recorded (Ban & Vrtiprah, 1999). On the other hand, in the summer of 2020, the first and the worst pandemic year, Dubrovnik recorded almost half a million visits and nearly 3 million overnight stays, as seen in Figure 16.4. These revelations also support claims that wars or armed conflicts could grip tourism destinations more severely than crises of other origins (Lachhab et al., 2022).

Stojčić et al. (2023) were interested in Dubrovnik's recovery trends, and their analysis showed that the first post-pandemic year, 2022, resulted in even more tremendous pressure on the Old Town than in the record-breaking pre-pandemic year 2019. They estimated that there were 3.448 daily visits to the historical core

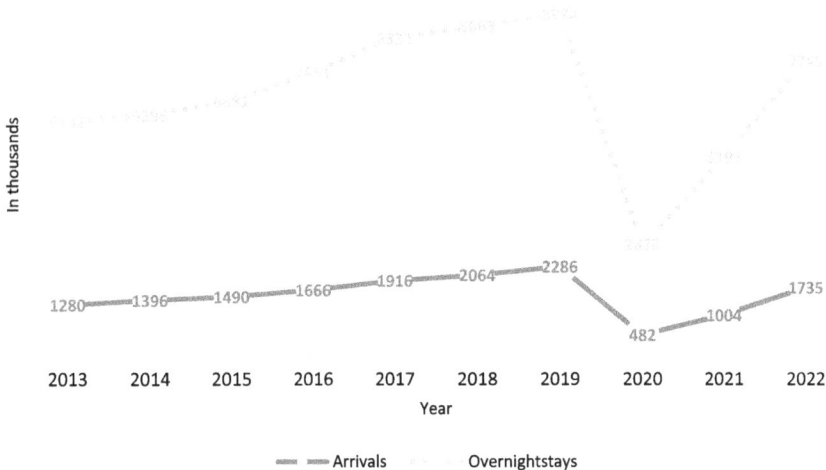

Figure 16.4 Tourist arrivals and overnight stays in the city of Dubrovnik, 2013–2022, in thousands

Source: Data collected from Visit Dubrovnik (n.d.b)

compared to 2.732 in 2019 and that there was a 25% increase in the overall number of visitors to the historical centre in 2022 despite the overall number of visitors to the city of Dubrovnik being 25% less than in 2019 (Stojčić et al., 2023). Hence, there were fewer visitors to the city of Dubrovnik but more visitors to the Old Town. Such trends do not bring fruitful results to some strategic objectives of the Respect the City initiative, not, for example, testifying well to increasing the quality of life of residents and preserving the cultural heritage (Stojčić et al., 2023).

However, these trends also need to be looked at in more detail – such as observing how visitor flow dispersed across different months. Stojčić et al. (2023) analysed that the months leading to Durbovnik's summer season recorded higher visitations post-pandemic than in the years before. The same could be observed for the months after Dubrovnik's summer season. Hence, there were more visitors in June and September post-pandemic than before the pandemic. Such analysis shows that despite more visitors in the Old Town, these were better dispersed pre- and post-season, creating less crowding and congestion. In this sense, the Respect the City initiative is actually proving to bring positive results, especially when promoting the dispersal of visitors within the city and beyond and stimulating new itineraries and attractions to boost out-of-high-season visitations.

We also have to reflect on the activities of Dubrovnik's tourist board that evolve around destination marketing. Significant efforts were put forward in the city's promotion linked to film-induced tourism. Šegota (2018b) observed that Dubrovnik was crowned the king of *Game of Thrones,* resulting from excessive social media promotion linked to the popular HBO series. In the pre-pandemic years, Dubrovnik also found itself on the infamous tourism destinations avoidance list due to excessive crowding and overtourism (Camatti et al., 2020; Mandić et al., 2023). Hence, the tourist board faced difficulties finding the 'just right' promotion to avoid overtourism and keep the visitors interested in the city. Of course, all promotional activities stopped with the pandemic. However, faced with significant losses in terms of visitor numbers, the tourist board found a new way of boosting Dubrovnik's popularity, which may go against RTC's strategic objectives. For example, while RTC focuses on limiting one-day visitors from cruise ships to the city, its tourist board is boasting about Dubrovnik winning Europe's Leading Cruise Destination 2021 and 2023 awards (Dubrovnik Tourist Board, 2023a). The latter feels as if Dubrovnik initially bit the hand that fed it before the pandemic and very soon bit its tongue when it was hit the hardest among Croatian tourist destinations during the pandemic.

On the other hand, as part of the RTC, a few stakeholders decided to tackle irresponsible visitor behaviour, also exhibited by cruise tourists. Hence, the city of Dubrovnik and its tourist board created an animated video to visually and entertainingly demonstrate examples of behaviour that aligns with the city's rules. Partnering further with Cruise Lines International Association and Croatia Airlines, the video was aired on all flights and cruise ships during the voyages that made stops in Dubrovnik (Dubrovnik Tourist Board, 2023b). At the moment of writing this chapter, there is no evidence of the impact of the animated video, so we have yet to see its results.

Conclusions

Finding the 'just right' balance between visitor satisfaction, resident quality of life, and conservation and protection of natural and cultural heritage could be a 'mission impossible'. Topping that with managing a tourist destination proclaimed 'extraordinary' on many levels adds to the impossibility. However, small steps in participation and cooperation of all stakeholders, who are willing to acknowledge their role and responsibilities in destination management, can yield great results.

The city of Dubrovnik, crowned the Pearl of the Adriatic, listed as a UNESCO Heritage Site, awarded the leading European Cruise Destination, and proclaimed King's Landing, found itself enjoying the perks and perils of all these titles. Its tourism industry boasted record-breaking numbers, but it also experienced significant negative tourism impacts, resulting in major losses of locals in the Old Town and visitor dissatisfaction. As a response, the Respect the City initiative was developed, focusing on integrated destination management through short-term, medium-term, and long-term measures and activities. RTC saw many stakeholders take responsibility for making Dubrovnik's tourism more sustainable.

The initial assessments of RTC activities show positive results towards taking Dubrovnik on a journey from overtourism to sustainable tourism. For example, the most perilous issues of cruise tourism and crowding have been addressed and positive results are evident in the dispersing of visitors across space and time. However, there is still room for improvement. As observed, Dubrovnik will need to emphasise rethinking its promotional activities so that positive short- and medium-term results are not overrun by the quest for popularity. The only popularity the city should strive for is being the leader of sustainable and responsible tourism development in the Mediterranean, a long-term goal that demands coordinated, consensus-driven, and accountable cooperation of all stakeholders.

References

Arezki, R., Cherif, R., & Piotrowski, J. (2009). *Tourism Specialization and Economic Development: Evidence from the UNESCO World Heritage List* (No. 176, IMF Working Paper).

Arslan, G., & Allen, K.-A. (2022). Exploring the association between coronavirus stress, meaning in life, psychological flexibility, and subjective well-being. *Psychology, Health & Medicine*, *27*(4), 803–814. https://doi.org/10.1080/13548506.2021.1876892.

Ban, I., & Vrtiprah, V. (1999). Turizam U Dubrovniku Pod utjecajem rata / The influence of war on tourism in Dubrovnik. *Acta Turistica*, *11*(2), 226–246. http://www.jstor.org/stable/23234055.

Beeton, S. (2010). The advance of film tourism. *Tourism and Hospitality Planning & Development*, *7*(1), 1–6. https://doi.org/10.1080/14790530903522572.

Bolan, P., & Williams, L. (2008). The role of image in service promotion: Focusing on the influence of film on consumer choice within tourism. *International Journal of Consumer Studies*, *32*, 382–390. https://doi.org/10.1111/j.1470-6431.2008.00672.x.

Bresson, G., & Logossah, K. (2011). Crowding-out effects of cruise tourism on stay-over tourism in the Caribbean: Non-parametric panel data evidence. *Tourism Economics*, *17*(1), 127–158. https://doi.org/10.5367/te.2011.0028.

Camatti, N., Bertocchi, D., Carić, H., & van der Borg, J. (2020). A digital response system to mitigate overtourism. The case of Dubrovnik. *Journal of Travel and Tourism Marketing*, *37*(8–9), 887–901. https://doi.org/10.1080/10548408.2020.1828230.

Cuccia, T., Guccio, C., & Rizzo, I. (2017). UNESCO sites and performance trend of Italian regional tourism destinations. *Tourism Economics*, *23*(2), 316–342. https://doi.org/10.1177/1354816616656266.

Currie, R. R., Wesley, F., & Sutherland, P. (2008). Going where the Joneses go: Understanding how others influence travel decision-making. *International Journal of Culture, Tourism and Hospitality Research*, *2*, 12–24. https://doi.org/10.1108/17506180810856112.

de Fauconberg, A., Berthon, P., & Berthon, J. P. (2017). Rethinking the marketing of World Heritage Sites: Giving the past a sustainable future. *Journal of Public Affairs*, 1–7. https://doi.org/10.1002/pa.1655.

Del Chiappa, G., Lorenzo-Romero, C., & Gallarza, M. (2018). Host community perceptions of cruise tourism in a homeport: A cluster analysis. *Journal of Destination Marketing and Management*, *7*, 170–181. https://doi.org/10.1016/j.jdmm.2016.08.011.

Dubrovnik Tourist Board (n.d.). Respect the City. https://tzdubrovnik.hr/lang/en/get/kultura_i_povijest/75283/respect_the_city.html#.

Dubrovnik Tourist Board (2023a). Dubrovnik odlikovan dvama prestižnim nagradama na World Travel Awards 2023. https://tzdubrovnik.hr/get/vijesti/83048/dubrovnik_odlikovan_dvama_prestiznim_nagradama_na_world_travel_awards_2023.html.

Dubrovnik Tourist Board. (2023b) The city of Dubrovnik and the Dubrovnik Tourist Board present a new animated film 'Respect the City'. https://tzdubrovnik.hr/lang/en/get/vijesti/82591/the_city_of_dubrovnik_and_the_dubrovnik_tourist_board_present_a_new_animated_film_respect_the_city.html.

Dubrovnik Visitor (n.d.). https://dubrovnik-visitors.hr/.

Frey, B. S., & Steiner, L. (2013). World Heritage List. In I. Rizzo & A. Mignosa (Eds.), *Handbook on the economics of cultural heritage* (pp. 171–186). Edward Elgar.

HBO (2016). http://www.makinggameofthrones.com/production-diary//game-of-thrones-triumphs-2016-emmys.

Huang, C., Wang, Y., Li, X., Ren, L., Zhao, J., Hu, Y., Zhang, L., Fan, G., Xu, J., Gu, X., Cheng, Z., Yu, T., Xia, J., Wei, Y., Wu, W., Xie, X., Yin, W., Li, H., Liu, M., et al. (2020). Clinical features of patients infected with 2019 novel coronavirus in Wuhan, China. *The Lancet*, *395*(10223), 497–506. https://doi.org/10.1016/S0140-6736(20)30183-5.

Jordan, E. J. (2014). Planning as a coping response to proposed tourism development. *Journal of Travel Research*, *54*(3), 316–328. https://doi.org/10.1177/0047287513517425.

Karagöz, D., Suess-Raeisinafchi, C., Işık, C., Dogru, T., Šegota, T., Youssef, O., Rehman, A., Ahmad, M., & Alvarado, R. (2023). Event motivation, subjective well-being, and revisit intentions during the second wave of the pandemic: Moderating effect of affective risk about COVID-19 and perceived trust. *Current Issues in Tourism*, *26*(24), 4069–4086. https://doi.org/10.1080/13683500.2022.2158787.

Kobašić, A. (1993). Turizam u razvoju dubrovačkog gospodarstva tijekom XX. stoljeća. *Ekonomska Misao i Praksa*, *2*(2), 97–115.

Lachhab, S., Šegota, T., Morrison, A. M., & Coca-Stefaniak, J. A. (2022). Crisis management and resilience – The case of small businesses in tourism. In M. E. Korstanje, H. Seraphin, & S. W. Maingi (Eds.), *Tourism through troubled times* (pp. 251–270). Emerald Publishing.

Larson, M., Lundberg, C., & Lexhagen, M. (2013). Thirsting for vampire tourism: Developing pop culture destinations. *Journal of Destination Marketing and Management*, *2*(2), 74–84. https://doi.org/10.1016/j.jdmm.2013.03.004.

Li, J., Nguyen, T. H. H., & Coca-Stefaniak, J. A. (2021). Coronavirus impacts on post-pandemic planned travel behaviours. *Annals of Tourism Research*, *86*, 102964. https://doi.org/10.1016/j.annals.2020.102964.

Li, S., Li, H., Song, H., Lundberg, C., & Shen, S. (2016). The economic impact of film tourism : The case of the *Lord of the Rings* and *Hobbit*. *Tourism Travel and Research Association: Advancing Tourism Research Globally.* http://scholarworks.umass.edu/ttra/2016/Academic_Papers_Visual/7.

Liberatore, G., Biagioni, P., Ciappei, C., & Francini, C. (2022). Dealing with uncertainty, from overtourism to overcapacity: A decision support model for art cities: The case of UNESCO WHCC of Florence. *Current Issues in Tourism*, *26*(7), 1067–1081. https://doi.org/10.1080/13683500.2022.2046712.

Lisle, D. (2016). *Holidays in the danger zone: Entanglements of war and tourism*. University of Minnesota Press.

Lowenthal, D. (1985). *The past is a foreign country*. Cambridge University Press.

Mandić, A., Pavlić, I., Puh, B., & Séraphin, H. (2023). Children and overtourism: A cognitive neuroscience experiment to reflect on exposure and behavioural consequences. *Journal of Sustainable Tourism*, 1–28. https://doi.org/10.1080/09669582.2023.2278023.

Marcotte, P., & Bourdeau, L. (2012). Is the World Heritage label used as a promotional argument for sustainable tourism? *Journal of Cultural Heritage Management and Sustainable Development*, *2*(1), 80–91. https://doi.org/10.1108/20441261211223289.

Martín, J. M. M., Martínez, J. M. G., & Fernández, J. A. S. (2018). An analysis of the factors behind the citizen's attitude of rejection towards tourism in a context of overtourism and economic dependence on this activity. *Sustainability (Switzerland)*, *10*(8). https://doi.org/10.3390/su10082851.

Mazanec, J. A., Wöber, K., & Zins, A. H. (2007). Tourism destination competitiveness: From definition to explanation? *Journal of Travel Research*, *46*(1), 86–95. https://doi.org/10.1177/0047287507302389.

Mihalič, T. (2020). Concpetualising overtourism: A sustainability approach. *Annals of Tourism Research*, *84*(July). https://doi.org/10.1016/j.annals.2020.103025.

Mihalič, T., Šegota, T., Knežević Cvelbar, L., & Kuščer, K. (2016). The influence of the political environment and destination governance on sustainable tourism development: A study of Bled, Slovenia. *Journal of Sustainable Tourism*, *24*(11), 1489–1505. https://doi.org/10.1080/09669582.2015.1134557.

Milano, C. (2017). *Overtourism and tourismphobia: Global trends and local contexts*. Ostelea School of Tourism & Hospitality.

Patuelli, R., Mussoni, M., & Candela, G. (2016). The effects of World Heritage Sites on domestic tourism: A spatial interaction model for Italy. In R. Patuelli & G. Arbia (Eds.), *Spatial econometric interaction modelling* (pp. 281–315). Springer International Publishing.

Paulauskaite, D., Powell, R., Coca-Stefaniak, J. A., & Morrison, A. M. (2017). Living like a local: Authentic tourism experiences and the sharing economy. *International Journal of Tourism Research*, *19*(6), 619–628. https://doi.org/10.1002/jtr.2134.

Plan Bleu (2022). *Catalogue of best practices on sustainable tourism in the Mediterranean.* https://www.dubrovnik.hr/uploads/posts/16025/KATALOG-NAJBOLJIH-PRAKSI-ODRZIVOG-TURIZMA-NA-MEDITERANU.pdf.

Puljić, I., Knežević, M., & Šegota, T. (2019). 'Overtourism'? Understanding and managing urban tourism growth beyond perceptions volume 2: Case studies. In World Tourism Organization et al. (Eds.), *'Overtourism'? Understanding and managing urban tourism growth beyond perceptions volume 2: Case studies* (pp. 40–43). UNWTO.

Šegota, T. (2018a). (G)A(i)ming at the throne: Social media and the use of visitor-generated content in destination branding. In C. Lundberg & V. Ziakas (Eds.), *Routledge handbook on popular culture and tourism* (pp. 427–438). Routledge.

Šegota, T. (2018b). Creating (extra)ordinary heritage through film-induced tourism: The case of Dubrovnik and *Game of Thrones*. In C. Palmer & J. Tivers (Eds.), *Creating heritage for tourism* (pp. 115–126). Routledge.

Sharma, G. D., Thomas, A., & Paul, J. (2021). Reviving tourism industry post-COVID-19: A resilience-based framework. *Tourism Management Perspectives*, *37*(October), 100786. https://doi.org/10.1016/j.tmp.2020.100786.

Siegel, L. A., Tussyadiah, I., & Scarles, C. (2023a). Cyber-physical traveler performances and Instagram travel photography as ideal impression management. *Current Issues in Tourism*, *26*(14), 2332–2356. https://doi.org/10.1080/13683500.2022.2086451.

Siegel, L. A., Tussyadiah, I., & Scarles, C. (2023b). Exploring behaviors of social media-induced tourists and the use of behavioral interventions as salient destination response strategy. *Journal of Destination Marketing and Management*, *27*(December), 100765. https://doi.org/10.1016/j.jdmm.2023.100765.

Sindik, J., Manojlović, N., & Klarić, M. (2017). Percipirani učinci kruzing turizma kod stanovnika Dubrovnika. *Ekonomska Misao i Praksa*, *1*, 151–170.

Stojčić, N., Pavlić, I., & Puh, B. (2023). *Analiza trendova posjeta staroj gradskoj jezgri Grada Dubrovnika temeljem podataka sustava Dubrovnik Visitor*. Report by Croation Development Agency.

Tkalec, M., Zilic, I., & Recher, V. (2017). The effect of film industry on tourism: *Game of Thrones* and Dubrovnik. *International Journal of Tourism Research*, *19*(6), 705–714. https://doi.org/10.1002/jtr.2142.

UNESCO (n.d.). Old city of Dubrovnik. http://whc.unesco.org/en/list/95/.

Urry, J., & Larsen, J. (2011). *The tourist gaze 3.0* (3rd edition). Sage.

Visit Dubrovnik (n.d.a). About Dubrovnik. http://visitdubrovnik.hr/index.php/en/26-uncategorised/cities-towns/570-about-dubrovnik-eng.

Visit Dubrovnik (n.d.b). Statistics. https://visitdubrovnik.hr/hr/o-nama/statistika/.

Vrtiprah, V., & Dragičević, M. (Eds.) (2017). *Marketing as a pillar of success-competitiveness, co-creation and collaboration: Book of abstracts*. University of Dubrovnik, Department of Economics and Business Economics

Part IV
Conclusion

17 Reflections and Future Perspectives on Tourism Interventions

Jeroen Nawijn, Jelena Farkić, Jeroen Klijs, and Rami K. Isaac

The contributors to this volume represent the diversity of interests examining the effects of various tourism interventions. The book takes a global perspective, exploring the diverse geographical contexts in which these interventions are tested, implemented, and analysed, ranging from the USA and European countries to Indonesia. From this point of view, we trust this volume is relevant not only for academic readers but also for policymakers and managers as they review their strategies, practices, and realities of tourism interventions and how they are utilized not only to optimize the impacts of tourism but also to explore how they can serve as productive ways of sustainable tourism development. In Part I, the concept and current state of interventions in tourism was discussed by multiple authors, while in Parts II and III, we presented a range of innovative tourism interventions. The chapters across all three parts pinpointed characteristics of interventions that have been and could be used to achieve future success in making, rather than breaking, destinations.

We hope that this volume will accelerate the growth of the studies exploring different types of interventions in tourism destinations. While this volume's contributors have ranged widely across topics, approaches, and places, it is worthwhile to summarize a few thematic elements in detail and highlight further developments and future trends. These themes, such as meaningful connections with(in) communities, the power of storytelling, and commitment and responsibility, reflect some of the contemporary issues in tourism generally, and we briefly recap them in this concluding chapter. They are, however, interwoven throughout the chapters, making neat separation between them impossible, and unnecessary. Being cognizant of their overlapping nature, we provide their more detailed discussion across the ensuing sections.

Meaningful Connections With(in) Communities

The notion of 'connection' is singled out as one of the key themes, reflecting the integral role of local engagement in creating authentic, meaningful, and mutually beneficial relationships within the tourism context. The concept of authenticity, and authentic experiences, has long been part of the discourse within the tourism academia and industry (Sharma & Rickly 2021; Wang 1999), while the notion of

DOI: 10.4324/9781003449027-21

meaningful experiences has gained traction more recently. More frequent discussions on, for example, well-being have helped deepen our understanding of the processes of making meaningful connections. With the emergence of the so-called 'positive tourism' scholarship exploring subjective, as well as local community well-being, new spaces have opened up for exploring how meaningful interactions unfold in diverse tourism spaces (Buzinde et al. 2014; Filep & Pearce 2013). Furthermore, the transformative turn in tourism has also yielded valuable insights into how individuals construct meaning in the ever-changing, and challenging, world (Pritchard et al. 2011). Sheldon (2020: 5) suggests that 'each personal interaction can create a moment of awakening, however, exchanges that are deep, intimate and soulful are more likely to have a long-term transformative effect'. Interventions that enable authentic connections may therefore be operationalized through various avenues, such as storytelling, creative engagements, or even challenging circumstances, where the compassion and understanding of both familiar individuals and strangers might enable the processes of connection.

Recent developments, however, catalysed by various global events, have brought a renewed focus on searching for and designing authentic and meaningful experiences in diverse tourism-scapes. On the one hand, local residents, as crucial stakeholders, seek to maintain a connection with the places in which they live, rather than alienate themselves from them due to the pressures of tourism. On the other hand, tourists seek encounters that allow for creating meaningful connections to the people and places they visit (Rickly 2022). Despite this shared striving for connection, authenticity, and meaning(fulness), the challenge remains to balance tourism development in such a way that it contributes to the satisfaction of all involved stakeholders. However, the active involvement of local residents in shaping destinations appears limited, except in situations where challenges like overtourism come into play (Liberatore et al. 2023). The shifting dynamics between residents and tourists therefore call for the growing importance of managing interactions and experiences in tourism places. It is especially the case in the Global South that local residents are often misrepresented as if they are mere hosts to guests (Stone & Nyaupane 2020). This is a pity as their agency and active involvement in policy-making and shaping a destination, for domestic and international visitors, is key to developing destinations sustainably. The engagement of residents therefore goes beyond the conventional role of hosts; it involves integrating their perspectives, situated knowledges, and contextual insights into the development of interventions that might impact their communities long-term.

To this end, tourism interventions presented across the book's three parts demonstrate a comprehensive effort to engage and benefit local communities. They highlight the urgency of prioritizing the local community and cultivating meaningful interactions with residents in the development of sustainable tourism. A community-centric approach is therefore crucial, as it primarily contributes to the authenticity and preservation of their heritage. Behavioural interventions, such as nudges, are discussed in Chapter 2 as mechanisms to influence individuals to make decisions that benefit society, the environment, and themselves. This implies a focus on encouraging responsible behaviour among tourists to minimize negative impacts on local communities. Further, interventions rooted in artistic research projects, as discussed in Chapter 3, also suggest an effort to engage the public in shared spaces, potentially enabling the development of a sense of belonging as well as community identity and pride. Physical interventions, exemplified by the Fair Isle Observatory in Chapter 4, present a tangible positive impact on the local community in that it contributes not only to economic sustainability but also to the social and cultural values of the island. Processual interventions, as seen in the discussion of UNESCO geoparks in Chapter 5, engage with the transformation of landscapes for tourism, while emphasizing the need to consider the well-being of local communities amidst the commodification processes.

Cutting-edge interventions in Part II, such as arts- and service design-based placemaking, storytelling in museums, and heritage walks, suggest the importance of integrating the stories and histories of the place into the local narratives. These interventions aim to present a critical view of traditional approaches, acknowledging and incorporating the stories and ways of being of local communities. The theme of connection is addressed rather often throughout this edited book. For instance, in Chapters 7 and 8, the authors observe that it is key to design interventions that trigger empathy and contribute to building the resilience of tourism destinations and their communities. The participation of locals as 'intercultural companions' in initiatives like Migrantour discussed in Chapter 8 further highlights the value placed on involving residents in the tourism experience. Living labs in the Netherlands, as discussed in Chapter 9, further emphasize the importance of collaboration with the community as an active participant in shaping the tourism landscape. The Belgian St Godelina's Abbey project, discussed in Chapter 10, serves as an exemplary case of citizen participation in future development scenarios, reflecting a commitment to respecting the place and creating value for local residents. Chapter 12 discussed how the impact study on off-highway vehicle recreation in Arizona focuses on measuring the economic impact on local businesses and employment, while highlighting the significance of going beyond the immediate economic impact to consider the well-being of local residents. Chapter 16 suggested how finding the balance between visitor satisfaction, resident quality of life, and conservation and protection of natural and cultural heritage might seem unattainable, however taking small steps in participation and meaningful collaboration of all stakeholders can yield great results.

The diverse community-centred interventions discussed across the three parts range from nudges and artistic initiatives to participative inquiries and economic impact assessments, however they collectively reflect a commitment to inclusivity, sustainability, and the well-being of the residents in tourism destinations. Acknowledging the importance of local perspectives, the interventions outlined here promote cultural sensitivity and ethical ways of developing tourism, which is paramount for the long-term sustainability and making, rather than breaking, of destinations.

The Power of Storytelling

Over the past couple of decades, researchers have been increasingly drawn to the importance of integrating stories into the development of tourism destinations (Moscardo 2020; Mossberg 2008; Saarinen 2004). Stories have long served as powerful tools for constructing the identity and image of organizations, products, and destinations, while in tourism, businesses are built around narratives, with hotels, restaurants, attractions, or events conceptualized around telling attractive and absorbing stories. Indeed, the power of storytelling has been also increasingly utilized as a key tool in placemaking (Calvi & Hover 2021). This influence is evident not only in designing or creating visitor experiences but also in understanding storytelling as a co-creative process. Visitors are actively engaged through sharing personal narratives of their visits, thus contributing to the overall tourism experience. Despite the increasing use of storytelling in tourism, a consistent and comprehensive approach is still elusive, leaving uncertainty about the inherent effectiveness of storytelling interventions at tourism destinations.

Several chapters in this book provide unique insights into the use of storytelling as an intervention in the process of making places. For example, in Chapter 7, Mitas and colleagues explored the use of storytelling to enhance both tourist and resident experiences, focusing on the experience of a museum exhibition. The findings indicated a steady increase in emotional arousal from the beginning of the story to the end within the four museum rooms where the exhibition took place. The interventions discussed by Björn and Miettinen in Chapter 6 teach us that stories enable participants to attach meanings to their embodied experiences. The creative interventions performed in the natural environments of Arctic Finland were based on the collection of artistic processes carried out through utilizing various media as a means to design new artistic ways of placemaking. The authors showed how interventions emerging from creative placemaking within tourism have the potential to deepen a sense of connection between locals and tourists and therefore broaden their natural and cultural knowledges. Such initiatives may greatly contribute to decolonizing natural heritage through exploring the multiple ways of interacting with the environment and non-humans, thus breaking down dichotomies such as human/non-human or nature/culture, and creating new avenues for more creative and performative tourism interventions.

The integration of storytelling in tourism emerges as a powerful tool with the potential to significantly enhance the quality of life within local communities. As evidenced by the research discussed, storytelling goes beyond constructing attractive narratives for businesses; it serves as a co-creative process, actively engaging visitors in sharing personal narratives. The chapters in this book shed light on how storytelling interventions, such as those explored in museum exhibitions or creative placemaking in Arctic Finland, contribute to emotional arousal, meaning attachment, and a deeper connection between locals and tourists.

Commitment and Responsibility

Interventions aim to make a societal impact. Fuelling a sense of responsibility is important in achieving involvement and eventual commitment of residents and other stakeholders in order to contribute to making, rather than breaking, a destination (Su et al. 2023; Walker & Moscardo 2016). Hence, this book joins the growing canon of studies advocating for a re-evaluation of the notion of success in tourism, emphasizing the need to assess tourism impacts not solely based on quantitative metrics like tourist arrivals and expenditure but also taking into account the effects on broader societal values, including well-being and resilience of destinations and their communities. However, it is imperative to be aware of potential challenges and complexities associated with facilitating this sense of responsibility. Questions may arise about the effectiveness of strategies in truly empowering communities, ensuring their active participation, and addressing power dynamics within the intervention processes. Thus, a careful examination of, and reflection on, the ethical implications and unintended consequences of such interventions is essential to establish the balance between positive impacts and unintended drawbacks. Rather often, the failures, frictions, and tensions that emerge in the process can act as foundational elements for designing more successful interventions: a vital dimension that we also wanted to pay attention in this volume.

When striving to make an impact, commitment is essential. It is key in navigating a path towards achieving this goal, as we observed in Chapter 9 authored by Gerritsma and Horgan. In exploring urban living labs for leisure and tourism, this study identifies key factors that can make or break tourism interventions. Successful innovation in the living lab environment requires individuals who are willing to take risks and learn from their failures – even if it means embracing uncertainty – while still being dedicated to finding a way to make a difference. This essentially involves a process of trial and error, where established systems and orthodoxies are broken down and rebuilt through iterative prototyping. In addition, Chapter 15 provided a critical assessment of policy concerns and policy interventions with direct tourist consequences, as well as policy difficulties produced by unregulated and market-driven tourism regulations. It offered a framework depicting policy failures, policy-action linkages, policy implementation gaps, and non-policies based on evidence gathered from various locations and circumstances in developed and developing countries. Chapter 16 discussed the importance of short-, medium-, and long-term actions and activities aimed at positioning Dubrovnik as the Mediterranean's leader in sustainable and responsible tourism, while stressing the importance of re-assessing the strategic initiatives and local and national tourism development (mal)practices.

Future Interventions

Concluding this edited volume we will, based on all chapters and the points addressed in the previous sections, formulate practical recommendations to be taken into account when developing future interventions that aim to contribute to 'making places'. This is obviously a simplification of all the nuanced discussions that took place before, but we still hope it can function as a valuable agenda for practitioners involved in developing interventions.

When designed well, interventions in the tourism ecosystem contribute to meaningful and mutually beneficial interactions, relationships, and connections between tourists and residents. Active involvement of local residents in the development, implementation, and evaluation of interventions, integrating their perspective, situated knowledge, and contextual insights, is key for developing tourism experiences that are ethical, authentic, meaningful (for both tourists and residents), and (long-term) sustainable.

Interventions should positively impact the local community, contributing to resilience and individual and collective quality of life. Such a community-centric approach is crucial, guaranteeing that residents both benefit from tourism and are willing to engage and contribute to creating memorable tourist experiences.

Although the focus in this edited volume has been on interventions aimed at places and situations where tourists and residents (physically) come together and are aware of each other's presence, there needs to be awareness that not all tourists and residents will engage in these kinds of 'conscious' and 'direct' interactions. In some situations, residents do not or barely come across tourists and/or might not even be aware of the presence of tourists and the impacts they have on the community and their quality of life. Furthermore, some tourists might be less interested in engaging in conversation or other interactions with residents. However, through stories experiences could be co-created that will be valuable for both tourists and residents, deepening the connections between them and facilitating co-creative processes towards meaningful change. Still, the impacts of interventions on these residents and tourists also need to be considered and some interventions, largely beyond the scope of this edited volume, might even be specifically directed at them. An example of the latter might be a change in the tourism tax, used to fund investment in facilities that benefit the entire local community.

The success of interventions needs to be evaluated based on broad societal values. Most notably, the focus needs to be on their contribution to resilience and quality of life. Here it is imperative to take into consideration *all* impacts of the interventions, looking beyond the impacts on people consciously and directly engaged in resident-tourist interactions (as discussed above) and beyond intended outcomes.

As will be apparent from the chapters in this book, interventions come in many shapes and sizes. They can, for example, be physical, based on events, stories and/ or art, legal, financial, or based on smart (IT) implementations. They differ in their intended purpose, e.g., creating meaningful experiences, reducing emissions, and/ or contributing to income and employment. They differ in the amount of uncertainty and risk they pose, the funding they need, the stakeholders required to be involved, as well as the necessary level of support by and cooperation between these stakeholders. Their results are also measured and evaluated in different ways. This implies that the choice of which intervention to apply needs to be based on a careful consideration of the different options, knowledge of interventions' characteristics, pros and cons, and success conditions – and especially an appreciation of the specific local context in which they are employed.

Still, even after making careful decisions, developing successful interventions is – to some degree – a matter of trial and error. It remains difficult to predict how interventions will work out in practice. Acceptance of the uncertainty and risks, a willingness to learn, and a long-term commitment to 'make better places' are essential for the successful, and sustainable, development of tourism.

The editors believe that this volume depicts the patchwork of contemporary topics in the expanding area of tourism interventions. Despite its internal logic and many areas of agreement, this volume does not supply ready-made solutions or universally approved cutting-edge interventions or definitions. Some questions remain unanswered. We anticipate that this volume will serve as a kaleidoscope of perspectives on the dynamics of tourism and interventions, as scholars discover patterns, linkages, and paradigms in their work, and then go on to sharpen the ideas, practices, and stories in this field. We hope that the examples featured in this volume will inspire further investigation and motivate readers to contribute to the continued understanding of the practical value of tourism interventions, and their meanings across diverse theoretical contexts and geographical settings.

References

Buzinde, C. N., Kalavar, J. M., & Melubo, K. (2014). Tourism and community well-being: The case of the Maasai in Tanzania. *Annals of Tourism Research, 44*, 20–35.

Calvi, L., & Hover, M. (2021). Storytelling for mythmaking in tourist destinations. *Leisure Sciences, 43*(6), 630–643.

Carnegie, E., & McCabe, S. (2008). Re-enactment events and tourism: Meaning, authenticity and identity. *Current Issues in Tourism, 11*(4), 349–368.

Filep, S., & Pearce, P. (Eds.) (2013). *Tourist experience and fulfilment: Insights from positive psychology*. Routledge.

Greene, D., Demeter, C., & Dolnicar, S. (2023). The comparative effectiveness of interventions aimed at making tourists behave in more environmentally sustainable ways: A meta-analysis. *Journal of Travel Research.* doi:10.1177/00472875231183701.

Liberatore, G., Biagioni, P., Ciappei, C., & Francini, C. (2023). Dealing with uncertainty, from overtourism to overcapacity: A decision support model for art cities: The Case of UNESCO WHCC of Florence. *Current Issues in Tourism, 26*(7), 1067–1081.

Moscardo, G. (2020). Stories and design in tourism. *Annals of Tourism Research, 83*, 102950.

Mossberg, L. (2008). Extraordinary experiences through storytelling. *Scandinavian Journal of Hospitality and Tourism, 8*(3), 195–210.

Pritchard, A., Morgan, N., & Ateljevic, I. (2011). Hopeful tourism: A new transformative perspective. *Annals of Tourism Research, 38*(3), 941–963.

Rickly, J. M. (2022). A review of authenticity research in tourism: Launching the *Annals of Tourism Research* Curated Collection on Authenticity. *Annals of Tourism Research, 92*, 103349.

Saarinen, J. (2004). 'Destinations in change': The transformation process of tourist destinations. *Tourist Studies, 4*(2), 161–179.

Sharma, N., & Rickly, J. M. (2021). 'The smell of death and the smell of life': Authenticity, anxiety and perceptions of death at Varanasi's cremation grounds. In D. Chhabra (Ed.), *Authenticity and authentication of heritage* (pp. 78–89). Routledge.

Sheldon, P. J. (2020). Designing tourism experiences for inner transformation. *Annals of Tourism Research, 83*, 102935.

Stone, L. S., & Nyaupane, G. P. (2020). Local residents' pride, tourists' playground: The misrepresentation and exclusion of local residents in tourism. *Current Issues in Tourism, 23*(11), 1426–1442.

Su, L., Yang, X., & Swanson, S. R. (2023). The influence of motive attributions for destination social responsibility on residents' empowerment and quality of life. *Journal of Travel Research*, *62*(8), 1737–1754.

Walker, K., & Moscardo, G. (2016). Moving beyond sense of place to care of place: The role of indigenous values and interpretation in promoting transformative change in tourists' place images and personal values. *Journal of Sustainable Tourism*, *24*(8–9), 1243–1261.

Wang, N. (1999). Rethinking authenticity in tourism experience. *Annals of Tourism Research*, *26*(2), 349–70.

Index

Note: **Bold** page numbers refer to tables; *italic* page numbers refer to figures.

For Product Safety Concerns and Information please contact our EU
representative GPSR@taylorandfrancis.com
Taylor & Francis Verlag GmbH, Kaufingerstraße 24, 80331 München, Germany